To Boldly Go Where No Book Has Gone Before

Luke O'Neill is a world-renowned scientist, author and professor of biochemistry at Trinity College Dublin. He has published over 350 major papers, has six major discoveries to his name and in 2016 was made a Fellow of the Royal Society – 'the equivalent of a lifetime achievement Oscar' (*Guardian*). A prominent figure in science, Luke appears regularly on TV and radio, has a weekly column in the *Irish Independent*, and in 2021 chaired the Royal Society Science Book Prize. How can Luke ensure that he appears in a history of science? Write one. This is that book.

To Boldly Go Where No Book Has Gone Before

A Joyous Journey Through All of Science

LUKE O'NEILL

PENGUIN BOOKS

PENGUIN BOOKS

UK | USA | Canada | Ireland | Australia
India | New Zealand | South Africa

Penguin Books is part of the Penguin Random House group of companies whose addresses can be found at global.penguinrandomhouse.com.

First published by Viking 2023
Published in Penguin Books 2024
001

Typeset by Jouve (UK), Milton Keynes
Printed and bound in Great Britain by Clays Ltd, Elcograf S.p.A.

The authorized representative in the EEA is Penguin Random House Ireland, Morrison Chambers, 32 Nassau Street, Dublin D02 YH68

A CIP catalogue record for this book is available from the British Library

ISBN: 978–0–241–99412–2

www.greenpenguin.co.uk

Penguin Random House is committed to a sustainable future for our business, our readers and our planet. This book is made from Forest Stewardship Council® certified paper.

To Andy Gearing, friend, fellow scientist, and deserving of an asterisk for equal contribution to this book

Contents

Introduction

When I was 12 years old, my parents gave me a chemistry set for Christmas. Eagerly, I took it to my bedroom, cleared the small table of the toy soldiers I had lined up, and my journey into science began. I laid out the test tubes and various chemicals that came with the kit, and – oh joy – there was also a small glass jar for the methylated spirits burner. Fire and chemicals – what a Christmas it promised to be!

Then I opened the little manual that came with the kit, and disappointment struck. A lot of boring writing. It explained that 'methylated spirits' is neat ethanol (the component of alcoholic drinks that makes you drunk) mixed with some bitter and toxic chemicals (to stop it being drunk). Each molecule of ethanol is composed of two carbon atoms, six hydrogens and one oxygen, and when it is exposed to a spark it can combine with oxygen in the air to form water (H_2O) and carbon dioxide (CO_2), liberating an excess of energy, mainly as heat. Even now, when I smell methylated spirits (I have to get my kicks somewhere), I am transported back to my bedroom and my chemistry set.

Ignoring the manual, I decided to do my first experiment. I grabbed one of the biggest test tubes and added a spatula-full of every chemical in the kit. I added some water and shook everything vigorously. But nothing much happened. A *little* bit of fizzing, but nowhere near as much fun as the fizzing I'd see when I put three whole tablespoons of my dad's Andrews Liver Salts into a glass of water. So, I hit upon an idea. What if I heated it up? I placed the test tube in its stand, put a cork bung in the top, set it over the burner's flame and waited. Three long minutes later . . . BANG!

My mother ran upstairs to find not what I thought a *real* scientist should look like – singed eyebrows, frazzled hair and goofy grin – but a faintly guilty, excited boy and a huge splat of brown chemicals on the ceiling. Still, I was inspired. And despite multiple coats of paint, the brown stain remained. Later, as I did my duller but safer chemistry homework, I would look up at the ceiling and daydream of the halcyon days when a boy could blow up his bedroom.

In recent years, in a turn of events none of my scientist colleagues could have foreseen, I became a household name. Irish television made a documentary about my life, and we returned to film in that very bedroom nearly 50 years later. And guess what? The stain was still there! If only I could remember the concoction, I could patent some everlasting brown paint.

After the incident with the stain, my next scientific revelation came in a science lesson at secondary school. Our teacher, Gutty – he had that wonderful nickname because he loved nothing better than gutting all sorts of creatures in our biology lessons – stood with great showmanship at the front of the class and said, 'Now watch this, boys.'

He had two big test tubes, both half-filled with what looked like water. He poured one of these clear liquids into the other and it turned bright yellow, like egg yolk. I was amazed. How could two colourless liquids be so transformed? I learned that what he had mixed was lead nitrate and potassium iodide. The yellow precipitate that formed is called lead iodide. As Walter White in *Breaking Bad* (the number-one show for us chemistry nerds) said to his disinterested pupils, 'Chemistry is the study of change.' You see, a swapping of sorts had happened: the iodine moved from the potassium to the lead, forming a new chemical with totally different properties.

In another inspirational lesson with Gutty, he had six expensive and fragile mercury thermometers – one for each of the groups he divided us into. In a deep and serious voice, he said, 'Now listen, boys. If anyone in any group breaks a thermometer, I will punish all six boys in that group.'

The practical began, with our group using the expensive thermometer to measure the change in temperature of water when we added a certain amount of water of another temperature. No Nobel Prize for guessing that the temperature changed. We even drew graphs – Gutty said it was important to draw graphs if you were a scientist. We took turns handling the thermometer carefully and it was all going very well. At the end of the session, I was left with an improved understanding of energy transfer, and I was also left holding the thermometer. My job was to put it back in its plastic sheath – which, carefully, I did. It slid in nicely. However, I hadn't put a bung in the other end of the sheath.

I watched in horror as the thermometer slipped all the way through and smashed on the tiled laboratory floor, the mercury forming little silver balls that flashed across the room. I am not sure it is scientifically accurate to say that time slowed down, but it sure felt like it did, and the whole class had gone silent. Lifting my eyes from the crime scene, I felt dozens of eyes on me. I translated the aghast stares of my five teammates as 'What on Earth have you done, O'Neill?' Slowly, Gutty walked over to our group and thundered, 'Two hours' detention at the end of the day to you all, during which you will copy out the glossary of terms at the end of the science book!' This was probably my first step in learning complex scientific terms – so thanks, Gutty.

As we left the classroom, I waited for the inevitable roughing-up from my fellows. But to my amazement, they came over to me, and one of them put his arm around my shoulders and said, 'Don't worry, Luke, it could have happened to any of us – you f###ing idiot.' We all had a good laugh and I felt for the first time the camaraderie among scientists that I have come to know well over the years.

I could have ended up in any one of several branches of science. Thirty kilometres south of Dublin – and near where I grew up – are the Wicklow Mountains. It's not an area known for its balmy weather, but I was fascinated to learn that 11,000 years ago it had been covered in glaciers. To learn more about the wonders of our changing world throughout time, I thought that perhaps at university I should study geology. But, you see, I also had a strange fascination with aquatic mammals like manatees, so I thought maybe zoology would be better. And then I wondered, what about organic chemistry? Not only was it interesting to learn about the molecules made of carbon that formed the basis for life, but these molecules often *smelled nice*, too.

I imagine that these reasons for wanting to dedicate my life to a particular topic were as random as anyone else's. Why does anyone decide to do anything, really? Despite all the science I have learned, I do not know the answer. In the end, I chose biochemistry; I am an immunologist now, and I love it.

When it comes to science, what got me as a child continues to inspire me. Science is all about understanding the world and how it works, and I am endlessly curious. I'm not alone in this. Most children want to understand the world and how it works, and all the questions they ask reveal them to be little scientists. Sadly, some children are put off science for ever. Just like Gutty, there are amazing, passionate science teachers in every country on this planet, and yet some children aren't as lucky as me – their introduction to science is less about fun experiments and more the sort of boring explanations of chemical reactions that I found in my chemistry set manual. And that's why I am writing this book.

This book is for everyone: those who already love science, and those who don't (at least not yet). Whichever camp you're in, I want to bring you back to that childlike sense of wonder you felt when you first heard about dinosaurs or dolphins or the double helix.

For me, the science bug really bit when as a 20-year-old student I did proper research for the first time. I was looking into an inflammatory disease called Crohn's disease, which affects the digestive system. If left untreated, it is debilitating, involving a lot of pain, ulceration and bleeding. It can be fatal, and often requires surgery. Researching what might be going wrong in the body for those suffering from Crohn's disease struck me as *very* interesting. I realized I might find out something new about the disease, and that knowledge could help make people's lives better.

This led me to a PhD in London, where I did more work on inflammatory diseases – this time rheumatoid arthritis, which is where the joints become painful. I'd cycle over from my lab in Lincoln's Inn Fields to St Thomas' Hospital to collect samples taken from patients, and then grow cells from the samples to try to find out what was going wrong in the diseased tissue.

One day I was allowed to watch the surgeon remove the joint tissue from a woman who had kindly volunteered part of her hip for my research. I met her after the operation to thank her and she asked me what I worked on, and I told her, 'Rheumatoid arthritis.' 'Oh,' she said, 'do you work on other

Eyetises?' I said I didn't know what she meant. She replied, 'Eyetises, do you work on other Eyetises?' I asked her again for clarification. She began to get annoyed. 'Eyetises! Eyetises! Like arthr-eyetis or menig-eyetis or dermat-eyetis.'

'Ah,' I said, 'yes, Eyetises! Now I get it!' I was happy to explain to her that any disease ending in 'itis' means it's an inflammatory disease. Inflammation happens when you have an injury or infection and involves the affected area becoming red, swollen, hot and painful. These all happen in rheumatoid arthritis, but we still don't know why. It's a question I've been trying to answer for almost forty years.

In fact, I have spent my whole career working on Eyetises. More specifically, for my PhD I worked on what were then new-fangled things called cytokines, which are made by the immune system – the part of your body that defends you against infection from bacteria, viruses, parasites and fungi (this last one sounds a bit less scary than the others, but suffice to say, I am not talking about shiitake mushrooms). Cytokines are like messengers that sound the alarm and wake up the immune system's troops to fight the invader. But in inflammatory diseases like Crohn's disease or rheumatoid arthritis, they turn on our own tissues, causing the immune system to run riot there, as if there is an infection. Only, from what we currently know, there isn't. This is the reason why a disease like rheumatoid arthritis is called an autoimmune disease – from the Greek *autós*, meaning 'self'. It turns out that cytokines cause most of the symptoms of that disease – and many others – and stopping them helps patients a great deal.

Since my PhD, my research has led to discoveries about the immune system, and I have helped in the discovery of new medicines that are currently being tested in clinical trials.

And so the boy who almost blew up his bedroom and couldn't sheath a thermometer was made a Fellow of the Royal Society in 2016. The *Guardian* rather grandly calls this 'the equivalent of a lifetime achievement Oscar'.

Best of all, this meant I got to sign the Fellows' book, which has the signatures of all my great heroes in science that I'd read about since childhood. The signature of every scientist you've ever heard of is in that book. They all turned up in London to sign it, with a few exceptions who had good excuses: Sigmund Freud asked for the book to be brought to him as he was ill, and he died not long after he signed, and Winston Churchill was rather busy in 1941.

On the day I signed the book, there was one final challenge to overcome: I wasn't wearing a tie. I hate wearing ties; it might be one of the main reasons I became a scientist. I was the only one not wearing a tie among all that were being made an FRS that day. But luckily, I bumped into the librarian, who *was* wearing a tie. I asked him if I could borrow his and he gladly obliged. I had been saved from the embarrassment of not following the correct dress code by the librarian of the Royal Society – another example of scientific camaraderie.

One of the thrills of that day was being given permission by the now tie-less librarian to look at three signatures in the book (I don't know why they only let you look at three – time pressures, I suppose). Everyone always wants to look at Isaac Newton's signature, or Charles Darwin's. I did too, but first I asked the librarian for two other names.

Honor Fell was a scientist who had been director of the Strangeways Research Laboratory in Cambridge, where I carried out postdoctoral work (yes, that's right, Strangeways – better known in the UK as the name of a prison). She made important contributions to the science of cell biology, working out how to grow cells and tissues outside the body. One other thing I knew about her was that she had been a great supporter of scientists arriving as refugees from Europe in the 1930s. I had also heard that she insisted on bringing her pet ferret to her sister Barbara's wedding. Now that's cool. I looked up her signature, which she wrote in the big book when she became a Fellow in 1953. It was rare in those days for a woman to be made an FRS. There's more progress to be made, but mercifully, that at least has changed.

I also looked up Edward Conway. He was an Irish biologist who made important discoveries in the chemistry of living tissues, particularly muscle and blood. He was made a Fellow in 1947 and was the last Irish biologist before me to be made a Fellow. My elevation to FRS wasn't noticed much in Ireland, other than a friend of mine who told me that FRS stands for 'Former Research Scientist'. With friends like that . . .

As for the third name, well that had to be Charles Darwin, whose description of evolution explains how life works. More on him later.

In this book you will read about discoveries made by many Fellows of the Royal Society, and lots of other scientists too. When you read about some scientific breakthrough, always remember that the finding was made by a human – a living, breathing person, as flawed as you or me – usually working in a team with fellow scientists. Science has all the ambition, ego, rivalry and emotion of any other human activity, and it focuses on one aim: to get to the truth of the matter. That's what makes science so great.

Like all human pursuits, it can go wrong – we don't always get to the truth right away. When a scientific observation is made, other scientists will try to repeat the work in their own experiments, and then improve on it, making new discoveries. If something isn't reproducible experimentally, the observation falls by the wayside. At its best, what lasts, after much deliberation, rigour and effort, is the truth.

As well as the camaraderie and lack of ties, I was drawn to science because it is based on data and experiments. The first person to be so bold as to say that scientists should carry out experiments was Francis Bacon. He was born in London on 22 January 1561, and at the age of 12 was studying at the University of Cambridge. That sounds impressive – and he was – but that was the typical age for students at the time. Bacon's medieval peers believed that diseases arose from an imbalance of the body's four humours – phlegm, blood, yellow bile and black bile. This belief wasn't based on any evidence, but rather stemmed from the then 1,800-year-old writings of the Greek physician Hippocrates. In Bacon's book *Novum Organum*, published in 1620, he argued that we should conduct experiments to explore nature, and then draw conclusions from those experiments, which is the basis of science. Mind you, the terms 'science' and 'scientist' weren't used at that time. People who did science were called natural philosophers. The Cambridge professor William Whewell would coin the term 'scientist' in 1834, although he only came up with it in response to a challenge by the poet Samuel Taylor Coleridge. So, without poetry there would be no scientists . . .

Bacon felt his new method was likely to be reliable, as believing any hypothesis strongly enough might give rise to bias. He argued that, to test whether diseases were caused by an external influence rather than being just something innate, healthy people should be exposed to things like wetness or cold to see if that would cause illness. He also wrote that experiments should be repeated again and again to check the results. He was the first to argue for empiricism, believing that all learning can only come from experience and observation. Writing in 1862, the historian William Hepworth Dixon captured Bacon's influence on what happened next: 'Bacon's influence in the modern world is so great that every man who rides a train, sends a telegram, follows a steam plough, sits in an easy chair, crosses the channel or the Atlantic, eats a good dinner, enjoys a beautiful garden, or undergoes a painless surgical operation, owes him something.' I certainly think about him every time I follow a steam plough or eat a BLT.

In this book, I will focus on the discoveries that have held up, and the scientists who made them and how they did it. I'm here to be your guide, as best I can – bringing you the human stories behind the great scientific discoveries so far made by our species. As we will see, science often proceeds in leaps, where one scientist makes an important discovery and then there is rapid progress from the scientific community at large. One giant leap for humankind, if you will. When such a leap happens, it's difficult to remember a time when what has been discovered wasn't known. How could there have been a time when we knew nothing about germs, for example?

Or it can be two steps forward and one step back. For example, the discovery in the 1940s that some toxic chemicals can kill rapidly dividing cancer cells was a major advance showing that cancer could be treated. However, those same chemicals also kill normal cells, and it took many years of careful research and clinical trials to optimize the use of chemotherapy drugs so that they could treat certain tumours without killing patients.

Another way science happens is as a long process of debate among scientists until a consensus is reached, especially if this involves complicated things like climate change or vaccines.

But while this experimenting is all well and good, why do we even bother? What compels us to know things? Curiosity, of course. But other animals have similar sensory apparatus (or better) and a similar drive to investigate environments for danger, food, water, shelter, sex – and yet there is no cat Einstein devising new laws of physics to predict the behaviour of space and time (although after a random scroll online you would be forgiven for thinking that cats had invented the internet), and no international consortium of dogs is operating an orbiting space station (although the first animal to orbit the Earth or 'go for extraterrestrial walkies' was a Russian dog, Laika). Science happens because our curiosity finds itself paired with our amazing human intellect.

We are a clever species – or as rocker Ian Dury would have it, 'There ain't half been some clever b######s,' which just about sums up the history of science. We love to solve puzzles, and science is in essence one huge puzzle. A particular puzzle I grappled with at Strangeways involved a very important inflammatory protein called IL-1, which belongs to the family of proteins mentioned earlier: the cytokines. In the body's immune response, IL-1 is the person on the boat who sends up a flare as a signal of distress. The only trouble is it is an especially mischievous protein in several inflammatory diseases, including my old favourites arthritis and Crohn's disease. I knew that if I could find the actual flare being fired off by IL-1, we'd potentially be able to further help patients with these diseases. At that time, it was understood that it was a kind of alert to drive inflammation, but no one knew exactly what it was.

I was able to measure the flare being fired by IL-1 using a technique called a band shift. When I first heard about it, I thought it was something to do with the romantic activities of rock stars ('shift' is an Irish slang term for a snog). But the band shift was actually an experimental technique where a band (meaning a line) moved up a gel, and the upward movement meant that there was activity in the sample I had prepared, from cells to which I had added the IL-1. The band moving up the gel was somewhat like the flare being fired up into the sky. I was so deeply satisfied at solving this puzzle that I used to carry around the record of the band shift in my pocket and show it to people in pubs. 'There's that scientist again, showing us his shifty bands and telling us he's found a flare going off,' they'd say.

You see, while science is often seen as cold and calculating and quantitative, it always involves emotion. Bet you never thought 'science' and 'emotion' would appear in the same sentence. But something I learned when I began giving talks at conferences is that the audience won't remember what I tell them. I can show the best data ever, but most of them won't remember it. However, they *will* remember how they felt. This has informed the way I talk about science ever since. When I talk about science, I want people to share in the joy I feel talking about it.

I remember one occasion after I gave a talk in Oxford, an elderly woman came up to me. She was in her nineties and had worked as a lab technician for decades. She said she still liked to come in and hear about the latest science, and that she had shed a tear during my talk. I asked her why, and she told me I had spoken about a scientist she used to work for – none other than Hans Krebs, a famous biochemist we'll read about later. The next morning, she gave me a gift of some of the glassware that Krebs had used in his experiments. That made *me* shed a tear, and it has pride of place in my office now.

Tears? Joy? Science? What on Earth are you talking about, Luke? I don't know about you, but I certainly felt joy when, in November 2020, the pharmaceutical company Pfizer – in collaboration with BioNTech – announced spectacular results for their vaccine against COVID-19. Science isn't simply wonderful; it can bring huge practical benefits.

Beyond this remarkable recent breakthrough, the majority of advances that have helped our lives happened between 1920 and 1970, when most people in developed countries got access to clean water, flushing toilets, vaccines, antibiotics, electric lights, phones, fridges and washing machines. Since 1970, though, things have slowed down in terms of technologies benefiting humanity. As author Simon Kuper has written, today's giant companies have essentially brought us social media and a delivery company. Hardly in the same league as previous benefits. But he's optimistic – and as you'll see, so am I – that rapid advances will happen again soon, spurred on by recent crises, as is so often the case. The Second World War spurred the development of the flu vaccine, antibiotics, jet engines, radar and microwave technology, just as COVID-19 led to the global rollout of mRNA vaccine technologies. In our current world of so-called permacrisis, it is consoling to think some good may come of such awful events.

But don't worry, this book won't linger on the negative. Aside from the dramatic stuff that affects the health and well-being of the entire human race, science also makes me laugh. Would you believe me if I told you that scientists do brush their hair and their teeth, and are even capable of conversation that won't put off a prospective partner? (I have a wife, and no, I didn't try to impress her with my band shifts.) I remember on my wedding day, my newly minted father-in-law, Leslie Worrall, got up to make his speech. He mentioned how both Margaret and I were scientists (Margaret is a biochemist who works on cancer) and said he wasn't in our league at all. In fact, he said that as far as he was concerned, 'DNA' stood for '**D**on't **N**o **A**nything'. It got a big laugh, helped by all the guests being tanked up on methylated spirits.

This is all to say, I believe that science is a fundamental human activity, involving memory, knowledge, intellect, social interaction, humour – you can't beat a good laugh to negate the disappointment of a failed experiment! – and all our senses and emotions. I see science as a great comfort and friend in times of trouble and lousy strife. It banishes darkness and lets in the light in so many ways. Of course, this can bring its own problems, but that's not science's fault. It's humans exploiting science and technology that leads to unintended or even malign consequences – like new weapons, or medicines that aren't properly tested and which then harm people. That misuse is something we need to watch out for, but by and large, the system works.

Science describes a body of knowledge that is as true as it can be, because the evidence overwhelmingly points in one direction. It's not perfect. It can be biased at times, or even wrong. But it has this great knack of being self-correcting. Bogus conclusions rarely last long, because true scientists only seek one thing, and that is the truth. Newton's laws of physics stood the test of time, but his work on the philosopher's stone (which allegedly turned base metals into gold) didn't.

In this book I will take you on a journey through the history of science, but one seen through my eyes. I enjoy doing science, but I also love telling people about it. I've given talks to people from every walk of life – from schoolchildren to college students, beauticians to bar owners, lawyers to prisoners. It has been a privilege to bring science to them all.

I am a huge advocate for the importance of scientific literacy. Without widely accessible education, all the science we have at our disposal becomes sealed into an elite vacuum, available only to those in the know. And how else can we help our children interrogate a world full of misinformation and a future that will increasingly involve AI bots? Science is for everyone, so we can employ it to improve our own lives and those of our companions on this planet.

John Comenius is a hero of mine. In the early 1600s he introduced a lot of things that are commonplace now: teaching being gradual; a focus on logic rather than learning by rote; and education for women and the poor. Comenius was also partly responsible for the cause of every student's back pain: textbooks. In 1639, he published a book called *Pansophiae Prodromus*, in which he attempted to give an account of all human knowledge at that time, written in a way that could be grasped by children. This was very nearly feasible, as the amount of academic knowledge back then was almost manageable. Impossible now, of course, given how much we've learned since then – and continue to learn. Although that hasn't stopped me from giving it a go here.

This book will loosely follow the history of science itself, beginning with humans looking up at the sky and wondering about the stars and planets – a wonderment which gave rise to both mathematics and astronomy. From there, we will travel from alchemy to the invention of chemistry, and how this was later used to help us understand the basis for life. As we will see, these advances led to important work on the human body and diseases, and to efforts to understand the most challenging black box of them all: the mind. I will show how we have arrived at a point where technologies shape our every waking hour, and explore whether we are smart enough now, with enough technologies at our disposal, to deal with that most pressing of concerns: climate change. So that's to say, strap in, we're covering a lot of ground.

But this is no ordinary history of science. For one, I hope the ride is an enjoyable one, and unlike other books on science, mine doesn't mainly focus on the details of theoretical and practical advances, or the whole 'wonderful world of science' thing – important though those things are. I've always wanted to know about the people who came up with the ideas and did the experiments that shaped science – real people, with personalities, passions, quirks and faults (so many faults), who have inspired and entertained me.

I wear many hats. I am an internationally renowned scientist, author and educator, and I love to sing and play the guitar. To my friends, I'm better known as the dilettante Irish guy who loves a pint. I might struggle to sing and play the guitar in these pages, but along the way, I want to introduce you to the other aspects of my character. As for my research, it has resulted in advances in understanding how key aspects of our immune systems work. Whether I ever end up in a future history of science is uncertain – writing one is a safer way to guarantee inclusion.

Science shows that life began around 4 billion years ago, when the first cell evolved, and then it kept on evolving, giving rise to all life on Earth: from the tiniest bacteria, *Nanoarchaeum equitans*, which is only 200 nanometres in diameter – a thousand times smaller across than a human hair – to one of the biggest dinosaurs, *Australotitan cooperensis*, which was as tall as a bus and as long as a basketball court and whose fossil was only discovered in 2021. Like everyone, scientists love *Guinness World Records* – biggest, smallest, oldest, weirdest . . . and some, like me, simply love a pint of Guinness. There's a lot I could have included in this book, and I apologize up front about any omissions, but life is short, and I don't want the book to be too long. I'm no Comenius.

Still reading? Good. Get ready for a book about science, the likes of which you'll never have read before, full of stories and characters. Do you believe me? No? I propose an experiment: read the book and find out if what I am claiming is true. Why, you're a scientist already . . .

1. City of Stars

Science fiction writers have had such fun with space. *Star Wars*, *Close Encounters of the Third Kind* and *Alien* would be very different films if they were about soil science or engineering. I went to see *Alien* when I was 15, and despite going through the birth of my children – and watching Ireland beat Romania in a penalty shootout during the Italia 90 World Cup – it still holds the record for me spending the most time watching anything through my fingers.

Similarly, the only place to start the story of science is space. You see, science is about measuring and recording things, and the earliest astronomers were fantastic at doing that. And they put these measurements and recordings to good use, by doing things like making predictions about when it would be best to plant crops.

Space has always been available for everyone to marvel at every cloud-free night, and everybody loves it. We all know about the solar system and the Milky Way and the vastness of space. Some of us even know about Einstein and relativity and the space-time continuum. Nobody understands it, of course, not even physicists, as several have admitted to me – at least not in the way we understand, say, how a bicycle pump works.

Homo sapiens first emerged as a species 200,000 years ago, and we have been looking up at the stars for at least that long. So where to start? As good a place as any is a burial chamber at Newgrange in Ireland, constructed 5,200 years ago. This building is older than those other ancient constructions tied into the cosmos that we all know so well: the Pyramids of Giza in Egypt and Stonehenge in England (both of which date to around 4,500 years ago).

Newgrange is a circular, cavernous tomb, 85 metres wide and 13 metres high. Imagine you're inside that tomb. It is pitch-dark in there all year round – except for approximately 17 minutes. If you're standing inside in the depths of winter, cold and damp and worrying whether the warmth of summer will ever return, something magical happens on the morning of the Winter Solstice (21 or 22 December). At sunrise, the light from the sun enters the chamber through a slit in an outer wall. It moves along a passage and then illuminates the darkness of the interior, before receding, plunging the inner chamber into darkness again – a darkness that won't be relieved for another year.

Forget VR headsets, these people knew how to blow minds. Imagine what our ancestors must have thought of this light show? The first person to witness it in modern times was archaeologist Michael O'Kelly, who saw the chamber light up in 1967 on . . . you guessed it, 21 December. O'Kelly was so affected by this ancient light show that he visited it every year until his death, despite being wary of the legend that the sun god Dagda would hurl the roof on him for being there.

Newgrange was built with remarkable precision by a community of Stone Age farmers, who transported massive boulders from the Wicklow and Mourne mountains, which are both over 100 kilometres away. How they were transported is still a mystery, but it was probably by sea and then via the nearby River Boyne. Both mountain ranges are pretty magical themselves, having been cut out by glaciers in the last ice age, 10,000 years ago.

The people who built Newgrange were closer to that ice age than to us, and yet they knew exactly at which point on the horizon the sun would rise on the shortest day of the year. They were able to align the sunbeam with the slit in the wall of the massive burial chamber. It's still not known how they did this, as the slit is on top of a massive boulder that had to be put into position precisely. A huge feat of early engineering and measurement (to get the exact position on the horizon where the sun would rise on the solstice), but also of astronomy. They knew that the sun rose at a slightly different point each morning, moving along the horizon, stopping (hence *solstice*, which means 'sun stopping') and then moving back again. Those ancient people could measure all of that, simply by observing and recording what happened each year, over a number of years. Incredibly, 5,000 years ago, people in Ireland were doing science in the same way that my lab does it today. They were carefully recording observations of a natural phenomenon, making predictions based on their observations and, finally, testing their predictions in an experiment.

The kind of life those people lived is largely unknown. There is in fact no evidence of large-scale settlements near Newgrange, which either makes this even more impressive a feat or, depending on how open to wacky conspiracy theories you are, points to it having been built by aliens.

The significance of the burial chamber at Newgrange is also not known. It is likely to have been related to agriculture and the management of food stocks – the drawing-up of calendars based on the sun and moon was important for deciding when to plant and harvest crops. The Winter Solstice marks the turning of the year, back towards the light. At least that's one idea. If you plant crops at the wrong time, they won't grow.

It most likely had religious importance too. And I like to think there might have been joy in measuring something accurately and predicting an event that came true every year. Science can soothe the mind in a troubling and uncertain world.

The small interior of the chamber – big enough to fit just a few people – has carvings on the walls, including intricate circles. Bones and grave goods such as pendants, a bone chisel and pins made from bone have been found in the chamber. DNA analysis has revealed a man was buried there whose parents were close relatives, possibly brother and sister. Such inbreeding in other places has been associated with royal dynasties, such as the pharaohs of ancient Egypt, who married among themselves to keep the royal blood untainted by mortals. The ancients certainly believed in keeping things in the family. It's possible those buried in Newgrange were the elite, and the sun was important to the religious beliefs of the community, perhaps lighting up the remains of the dead in their honour. We don't know for sure, but Newgrange as a burial chamber for royalty is a reasonable conclusion based on the DNA evidence. We'll get on to DNA later – for now, that's getting ahead of ourselves.

It blows my mind to know that the sun still lights up the inside of a burial chamber built over 5,000 years ago, for only 17 minutes each year, on the Winter Solstice. That's more than 200 generations ago. It also amazes me that Newgrange might have been like the Egyptian pyramids, where the Irish equivalent of pharaohs were buried. Which gives me the perfect opportunity to tell you my (and probably the only) Irish/ancient Egyptian joke. People look at me strangely when I say that I recently saw my first Irish-Egyptian hieroglyph. It read, 'Made in Eejit'.

Astronomy can be viewed as the oldest of the sciences because it involves observation and measurement and making a record, which are important for all branches of science. And Newgrange is a good starting point for all of that. It is also the first example of a Big Science project, involving the construction of a massive circular device that detects solar radiation – take that, Large Hadron Collider, you're 5,000 years behind the times.

Better than Magic

Astronomy became tied up with all kinds of things, including mythology, religion and astrology. Given how curious and intelligent we are as a species, it's no wonder that we looked up at the sky and wondered about what was going on up there. Many ancient cultures observed the night sky and saw the stars and noticed how they seemed to shift their position over the course of a year. Then there was the moon, which was constantly changing, but once it was systematically observed, it was seen to behave in predictable ways. Early astronomers could also see how some of the bright objects up there moved more quickly – these were subsequently shown to be planets. And how things like shooting stars and comets occasionally flared across the sky.

These changes were sometimes seen to be related to events on Earth, including natural disasters or even the births of important people, although there isn't any actual link. One thing scientists are big on is that correlation does not imply causation. Two things might be correlated – for example, a comet and an earthquake. But that doesn't mean that the comet caused the earthquake. Correlation is never sufficient for a scientist. There must also be evidence for causation.

In Shakespeare's *Henry IV, Part I*, the Welsh wizard Owen Glendower says: 'At my nativity the front of heaven was full of fiery shapes, of burning cressets; and at my birth the frame and huge foundation of the Earth shaked like a coward.' Many cultures mention celestial changes as harbingers of important events. One reason for this might be that, given all the uncertainty and anxiety in our lives, and our tendency therefore to anticipate the worst-case scenario, it was good to know that something might predict events, and thereby relieve some of this uncertainty. We most likely evolved our nervous nature as a survival mechanism. It's probably one consequence of being smart – we can take in lots of information and work through multiple possible future scenarios, both positive and negative. So, we worry a lot, and the stars might help us worry less, or at least prepare us for the worst.

The same goes for magic, which is common to all cultures and involves things like predicting the future, or the casting of spells for good or evil. All borne out of a sense of uncertainty. As archaeologist Chris Gosden has argued, three great strands run through human history: science, religion and magic, with both religion and science emerging after magic. The shaman existed long before the priest or the scientist. It might seem strange to write about magic at all in a science book, but magic is the idea that we have a connection with the universe, and that the universe somehow responds to us – so maybe magic isn't that far from science after all.

Astronomy is science, but astrology is definitely in the realm of magic. Astrology was an early attempt to use the positions and movements of heavenly bodies to predict human affairs and earthly events. Despite its 3,000-year-old history, there is no scientific basis for astrology, and there never was. Take that from me, I'm a Gemini and we're all very intelligent.

Some of the earliest drawings depict stars. There is a 32,500-year-old mammoth tusk depicting the Orion constellation, and the Lascaux caves in France feature paintings from 17,000 years ago that recognizably portray stars. These predate writing by a long way, since the first evidence for that dates to 3,400 BCE in Sumer, which is near the Persian Gulf.

The Egyptians aligned their pyramids to the North Star, most likely because it was the only star that was constant in the night sky. They also revered Sirius, or the Dog Star, because when it rose in the east before sunrise it predicted the flooding of the Nile, a critical event for agriculture.

An ancient shrine built by a people called the Hittites in what is now Turkey 3,000 years ago has been shown to be a symbolic representation of the stars and planets, and it is the earliest evidence we have of solar and lunar calendars. The Hittites thought the world began in complete chaos but quickly became organized on three levels: the underworld, the Earth and the sky. In the rocks they depicted stars that went around the North Star but never went below the horizon.

The origin of mathematics can be traced back to early astronomy. The Babylonians, through careful observation, realized that the length of daylight varied each day. They were a people living some 3,900 years ago in the city of Babylon, which was in Mesopotamia (modern-day Iraq and parts of Syria). Interestingly, the Babylonians based their numerical system on 60, not 10 as we do, but they divided the day into 24 hours, with each hour containing 60 minutes and each minute 60 seconds – sound familiar? Babylonians were also able to predict an 18-year cycle of lunar eclipses, and observed the regular movement of the planets. Meanwhile, in India, astronomers were able to observe the same phenomena and to explain that some of the observed movements suggested that the Earth was spinning on its axis. The Chinese were busy looking up too, with records of this from at least 3,000 years ago, and much of their astronomy was for timekeeping. Arabic and Persian societies under Islam also did a lot of stargazing and measuring. In the late tenth century, the largest observatory up to that point was built near Tehran, Iran. This allowed the astronomer Abu-Mahmud al-Khujandi to calculate the tilt of the Earth's axis relative to the sun. And Arabic astronomers gave names to several stars and constellations, which we still use. Betelgeuse is in the constellation Orion and means 'the underarm of Orion', who we hope for Betelgeuse's sake is wearing deodorant.

One of the most influential figures in astronomy is Claudius Ptolemy, who lived in Alexandria in the second century CE. Ptolemy published astronomical tables to help predict the locations of the sun, moon and planets in the sky, and his planetary hypotheses put the Earth at the centre of a series of nested spheres ending at the stars, which he calculated would be at a distance of around 20,000 times the radius of the Earth. Ptolemy used mathematics as the basis of this model of the universe, and not theology or metaphysics, as he thought it was the only way to secure knowledge with certainty. He also wrote a book called *Optics* that established many principles that would later lead to the development of telescopes.

In Europe meanwhile, the English monk Bede of Jarrow published an important book in 725 called *The Reckoning of Time*, which allowed the clergy to calculate the date for Easter. Bede wrote: 'The Sunday following the full Moon which falls on or after the equinox will give the lawful Easter.' Christians were therefore also using the stars to govern one of their main religious days. And of course, in the Bible, the Three Wise Men from the East followed a moving star to find the birthplace of Jesus in Bethlehem.

One Small, the Other Far Away

An awful lot of stargazing with the naked eye – whcther for pleasure (no TV or smartphones back in those days), religious or practical reasons – had therefore happened by the time the telescope was invented in the early 1600s. If rudimentary astronomy began 200,000 years ago, it took another 199,600 years for the properly rigorous science of astronomy to advance. A long wait.

The invention of the telescope enabled us to see much further. As with most scientific endeavours, a new technique or instrument allowed us to go beyond our natural abilities and observe something that otherwise could not be seen.

The person who gets the credit is Hans Lippershey, a spectacle-maker in Middelburg, the Netherlands, who filed a patent on 2 October 1608 'for seeing things far away as if they were nearby'. This is a mercifully brief description of an invention. I wonder if the writers of the *Father Ted* comedy show read it, inspiring the scene where Ted holds up a toy cow and has to explain to Father Dougal that it's small, but that the actual cows are far away.

There are several accounts of how Lippershey came up with his invention. One involves him watching children playing with lenses made for spectacles, and the children telling him that they could make a faraway weather vane seem closer. Lippershey's 'Dutch Perspective Glass' – or as he called it, a *kijker* (looker) – had a threefold magnification caused by the curved glass lens bending light, making it look like the objects viewed were three times closer than they were.

While Lippershey gets the credit, it's likely that lots of people were working with lenses and had made telescopes by combining more than one lens to increase magnification. The also-rans in this case include another Dutchman, Zacharias Janssen, as well as Englishman Leonard Digges and Frenchman Juan Roget. It can be difficult to assign credit in most branches of science, although that doesn't stop people from trying. And once credit is assigned, it's usual for someone else to shout, 'Hey! That was me!'

One thing for sure happened next. Lots of people all over Europe began making telescopes and pointing them at the night sky to get a better look at the stars. It became the latest trendy thing to do. We humans, and especially scientists, love jumping on a bandwagon.

By the time Lippershey described his telescope in 1608, a lot of astronomy had already been done with the naked eye. But now, lift-off could happen. The invention of the telescope opened up the cosmos. And the scientist who gets the most credit for that lift-off is one of the most famous scientists of all time, Galileo Galilei. He was in Venice in 1609 when he heard of the Dutch Perspective Glass. He went back to Padua the next day and immediately got to work making his own telescope. He kept modifying it, with different lenses and tube lengths, and then headed back to Venice to present it to the then Doge, Leonardo Donato. This resulted in him securing an academic post at the University of Padua and a doubling of his salary. If only things were so simple nowadays.

Galileo's improvements on the Dutch telescope, including

the use of a convex lens as the objective and a concave lens at the eyepiece in a long tube, resulted in one that was twenty times more powerful. It's likely that this combination was arrived at by trial and error, but we now know it's all about the bending and focusing of light. Galileo then began making observations. By the end of the year, he had observed the moons of Jupiter, hills and valleys on the moon, and sunspots – dark patches on the surface of the sun caused by the sun's magnetic activity – using a projection method rather than direct observation (unlike Donald Trump, he didn't look directly at the sun – he wasn't that stupid).

We then get the name 'telescope', which was coined by the Greek poet Giovanni Demisiani on 14 April 1611, at a banquet held in Galileo's honour. I love the idea of this big fancy dinner to celebrate an invention, and then one of the guests comes up with a name for it and everyone there loves it, breaking into applause. The banquet was hosted by Prince Federico Cesi at the Accademia dei Lincei in Rome; it's still going (the Accademia, not the banquet), and is one of the world's oldest scientific institutions.

Galileo's observations led to him championing the work of Nicolaus Copernicus, who some 60 years earlier had said that the Earth went around the sun, as opposed to the other way around. This is an important thing for scientists to do – give credit for that which was done before. Today we call this citation. Greek, Indian and Arabic astronomers – for example, the Greek philosopher Aristarchus in 230 BCE – had proposed this idea before. Copernicus based his conclusions on their observations, mentioning some of the Arabic studies in his own work. Curiously, although it's seen as a foundational study for all of science, the work of Copernicus didn't have much proof. It was mainly based on trying to fit observations into a model where the Earth and planets went around the sun. His book *De Revolutionibus Orbium Coelestium* caused something of a stir when it came out in 1543. He was dying of what was then called apoplexy (a stroke) on the day a copy was brought to him, and he is said to have woken briefly from a coma, looked at his book and then died peacefully. I hope this doesn't happen to me with this book.

The discovery that the Earth goes around the sun is an example of what the late Lewis Wolpert, who was a great popularizer of science, calls the 'unnatural nature of science'. Our eyes tell us that the sun moves across the sky, so surely it is the sun that moves, not the Earth. And yet the science irrefutably tells us that this is not the case, and the final proof (were it to be needed) was watching it ourselves from space. It can still be difficult to convince people of things, as we saw in the COVID-19 pandemic when many refused to believe that vaccines were safer than the consequences of catching the disease – or even that the SARS-CoV-2 virus existed at all.

Apart from supporting Copernicus, Galileo's work was significant because it found that Jupiter had moons going around it. But more importantly, it showed that the sun itself rotated, based on watching the movement of sunspots.

Galileo only ever referred to himself by one name, not because he saw himself as a Renaissance Sting or Cher, but because that was standard practice at the time. He did have something of a wild lifestyle, fathering three children out of wedlock. Two of his daughters entered the convent of San Matteo – because, being illegitimate, they couldn't marry.

On 30 November 1609, Galileo pointed his telescope at the moon and what he saw astonished him. Others had already tried using a telescope to look at the moon, including

English mathematician Thomas Harriot four months before Galileo, reporting that all he could see was a 'strange spottedness'. Harriot certainly missed a beat there. Galileo described lunar mountains and craters. He even tried to calculate the height of the moon's mountains based on the shadows they cast. This was a contrast to what was thought at the time, where the moon was, in the words of Aristotle, 'an eternal pearl'. Many of Aristotle's opinions – for example that the sun went round the Earth – had held sway since 330 BCE.

When Galileo saw the moons of Jupiter he named them the Medicean Stars after his patron, Cosimo II de' Medici. Galileo was also one of the first to observe the planets Saturn and Neptune; he thought the rings around Saturn might be moons. He was the first to describe the Milky Way, realizing that it contained a huge number of stars, which were previously thought to be clouds. He listed many stars that weren't visible to the naked eye.

But then poor Galileo got into trouble with the Catholic Church. Although, this may have mainly been because he upset the Pope, who had originally been a friend but felt he was portrayed unfavourably as a simpleton (the character Simplicio) in Galileo's book *Dialogo*. Even scientists need to be careful who they upset. The Roman Inquisition, which investigated people who weren't following Roman Catholic law, investigated him in 1633 and concluded that heliocentrism was 'foolish, absurd and heretical', since it contradicted the Holy Scripture. Galileo was forced to recant or face torture, imprisonment or worse, and spent the rest of his life under house arrest. Legend has it that, after he had recanted, he mumbled, 'And yet it moves.' Good for you, Galileo! He was also ordered to recite the Seven Penitential Psalms once a week for three years, but fortunately – for him – his daughter Maria Celeste was allowed to do this for him. She must have been bored rigid.

Galileo wrote an important work of science while under house arrest called *Two New Sciences*, which was mainly about different aspects of physics. It would lead to Einstein calling him the 'father of modern physics'. After he died, Galileo was buried in the Basilica of Santa Croce in Florence, although three of his fingers and a tooth were removed, and his raised middle finger can be seen today in Florence's Museo Galileo. Even in death Galileo was making a point to the Inquisitors.

Despite his books being placed on the *Index Librorum Prohibitorum* (Index of Forbidden Books) by the Church, Galileo would not be forgotten. His work on mathematics, physics and astronomy places him among the key people who helped to establish science in Europe in the 1600s. In the nineteenth century, Protestants used the story of how he was forced to recant to attack the Catholic Church. Finally, in 1992, Pope John Paul II said the Church had made a mistake in condemning him, and in 2008 a plan was announced to put up a statue to him in the Vatican. For some reason, that statue has yet to be erected, and so the story goes on.

Shrinking France

Galileo had a major influence on so many of the scientists that came after him. One was Giovanni Cassini, an Italian astronomer and engineer who lived in Italy from 1625 to 1712. Rather unpromisingly Cassini started out as an astrologer and believed the sun moved around the Earth, but like all good scientists he changed his views based on the evidence gathered from his studies. He carried out much of his work in France and even measured the length of the country. France turned out to be shorter than had been thought, and his patron, Louis XIV (appropriately enough called the Sun King), said that Cassini had taken more land off him than he had won in all the wars he had fought.

Following in Galileo's footsteps, Cassini found moons – only in his case they were going around Saturn. He also estimated the distance between Mars and the Earth at about 140 million kilometres. The enormity of the solar system and then the Milky Way galaxy and finally the universe was bit by bit figured out by astronomers. People must have been amazed when they heard of these vast distances, especially as ancient people had thought the stars were bright lights studding a ceiling that wasn't too far away. We still can't fathom these enormous cosmic distances, with us just a teeny tiny speck in all that vastness.

Johannes Kepler was another astronomer inspired by Galileo. He was born on 27 December 1571 near Stuttgart in Germany. His father had the unusual job of mercenary and was killed in the Eighty Years' War in the Netherlands, which I suppose was an occupational hazard for a mercenary. His mother was an innkeeper, healer and herbalist – I guess like a sort of proto-wellness influencer – which came in handy because Kepler was a weak child who was often ill. When he was young, though, he amazed people with his mathematical abilities, which just shows you how strange and random great ability can be. At the age of six he saw the Great Comet of 1577. It was described by an Italian observer as 'the comet shimmering from a burning fire inside the dazzling cloud', and by Japanese observers as being as bright as the moon. It was visible all over the world. Kepler's mother took him to a high place to look at it, and it made a huge impression on him. As a student he took part in a debate on the heliocentric versus the Ptolemaic system (where the sun goes around the Earth; named after Ptolemy), siding with heliocentrism. They must have been great days for student debating societies.

Later, Kepler took a job with one of the most famous astronomers of the time, Tycho Brahe. Brahe was a Danish astronomer who had made the most detailed observations on the stars and planets and is known as the last naked-eye astronomer. Tycho also owned a tame elk that died after drinking too much beer and falling down the stairs. As a student he had lost part of his nose in a sword duel with his third cousin Manderup Parsberg over who was the better mathematician, or possibly over a shared love interest. I can't imagine today's students having a duel over maths – love, maybe, but not maths. He wore a prosthetic nose for the rest of his life, and in 2010 his body was exhumed to find out what the nose was made of. It turned out to be brass, but records state that on special occasions he wore a gold or silver nose. Whatever was true about his nose, he worked out that the moon went around the Earth but thought that the sun also went around the Earth. He was half-right, but he believed the authority of Holy Scripture was paramount, and the scripture said the Earth was at the centre of everything . . .

Tycho had a large number of observations of the stars and planets, which he guarded jealously. Once he realized how smart Kepler was, he gave him access. When Tycho died, Kepler became his successor as imperial mathematician to Emperor Rudolph II, and so his future was secure. His main obligation was to provide the emperor with his horoscope every day and also the horoscopes of other foreign leaders. As daft as it seems, a recent American president also regularly took advice from an astrologer on affairs of state – no, not that one, it was Ronald Reagan. As far as we know, Kepler didn't think much of astrology, although he would occasionally do horoscopes for family and friends. Like most scientists, he knew he had to be conscious of what his funder wanted, and he probably did the horoscopes through gritted teeth.

Kepler was able to dig into all of Tycho's data, and what he saw astonished him. There were records of unexpected shadows on the sun, the red colour of a total lunar eclipse, and unusual light around a solar eclipse. But his most important finding was based on the orbit of Mars. He concluded that it went around the sun in an ellipse rather than a regular circle. This is now known as the first law of planetary motion. It was a striking finding, as an elliptical orbit is a strange oval shape. Surely orbits would be circular? That just seems like common sense. And yet they weren't. (We'll come back to that when we get to Newton.) Yet again, the unnatural nature of science. All the planets including the Earth go around the sun in an elliptical orbit. This information wasn't directly useful for anything, of course – as in fact might be said of much of astronomy, even today. It was just deeply satisfying for Kepler.

For fun, in 1608, Kepler wrote *Somnium* (Latin for 'The Dream'), a fictional account of a trip to the moon. This would cause trouble, as in it he described how the mother of the narrator consulted with a demon to learn how to travel in space. This seems to have started a trial against his own mother on a charge of witchcraft (remember, she was a herbalist), but thankfully she was acquitted, otherwise she might well have been drowned as a witch. Kepler later republished the book with 223 footnotes (which exceeded the length of the actual text), explaining how the book was an allegory and had considerable scientific content. I wonder what his poor mother made of all this? Not to mention the readers of the book (if there were many) who had to wade through all those footnotes.

That book goes down in history as the first work of science fiction, not an insignificant boast for Kepler. In his story, the demons push people 50,000 miles to the moon in four hours – a speed of 12,500 miles per hour. Not a bad prediction for the 1620s, as fast-forward to 1969 and the Apollo 11 spacecraft travelled the 239,000 miles to the moon at a speed of 25,000 miles per hour.

Kepler's life was to become even more difficult. His patron, Rudolph, was forced to abdicate by his brother Matthias. Both brothers asked Kepler for their horoscopes and wisely Kepler kept it all a bit vague and comforting. Galileo heard of his difficulties and recommended Kepler for his post in Padua, but he ended up in Linz in Austria, retaining the post of imperial mathematician. His wife had died, and at the age of 42 he remarried the 24-year-old Susanna Reuttinger, having considered 11 different matches over the previous two years. Who needs Tinder? He was obviously a man in demand.

Following his death, his books were widely read and hugely influential. Particularly on one young mathematician, Isaac Newton.

Unweaving the Rainbow

Isaac Newton was born on 25 December 1642, which like everyone born on that day must have been annoying to have as a birthday, as it meant fewer gifts. His father died three months before his birth, and he was born weighing three pounds, so he was probably around ten weeks premature. He was such a small baby that his mother said that he could fit inside a quart mug, which can hold almost a litre. His mother was keen that Isaac become a farmer, but his early brilliance was spotted by teachers. Again, the randomness of genius. He spent a lot of time as a teenager making sundials and models of windmills, and loved mathematics. At Cambridge University he read the works of Galileo and Kepler, which were very new and exciting at that time.

Because of an outbreak of the plague, he had to leave Cambridge for a time and return home. Those two years were perhaps the most productive period for any mathematician ever. He developed his theory of calculus, which is the study of continuous change. Newton had realized that simple algebra could be used to describe and predict the behaviour of, for example, an object moving at a constant speed in a particular direction, but it was useless in describing a falling ball that was accelerating under gravity or a planet constantly changing direction when orbiting a star. Calculus allowed him to understand these more complex properties of moving objects.

From his work on optics (the study of light) he showed that white light was composed of a mixture of other hues, outlined the basis of refraction (the bending of light by lenses), and even proposed that light behaved as though it were corpuscular (made of tiny particles). In his famous law of gravitation he provided a mathematical relationship for the force of attraction between two bodies – sadly, this does not refer to human attraction. His triple whammy was said to 'advance every branch of mathematics'. But many find the mathematics and physics of Newton difficult. A waggish student once said that everything would be much easier if the tree instead of the apple had fallen on Newton's head.

At Cambridge he was elected a Fellow of Trinity, which

meant becoming a priest. Newton refused and got a special exemption from King Charles II because of his fame as a mathematician. This is when he discovered that white light is made up of a spectrum of colours which split when it shines through a prism. For an illustration, check out the cover of Pink Floyd's *Dark Side of the Moon*, which also has its origins in Cambridge, since three members – Syd Barrett, Roger Waters and David Gilmour – grew up there. Have a listen too – it's a great album.

Newton discovered this by darkening his room but then allowing a beam of sunlight in, which he shone through a prism that he had bought at a nearby fair. This resulted in a rainbow being made in the room. Others had seen this before, but Newton showed that the multicoloured spectrum of light could be reassembled back into white light if the spectrum was shone through another prism. No one else seems to have thought of that experiment – or if they did, they didn't tell anyone. Somewhat sourly, the poet Keats accused him of unweaving the rainbow, and the artist William Blake portrayed him as focused on a scientific drawing implement while ignoring the beauty of nature. Art versus science – a battle as old as time . . .

In Cambridge he continued his work on gravitation, but turned his attention to the planets. Kepler's work intrigued him but so did the appearance in the winter of 1680–81 of a comet. Newton came up with the laws of motion, which describe the relationship between the motion of an object and the forces acting on it. He used them to explain and investigate the motion of many physical objects – and in particular, why planets have an elliptical orbit. He illustrated these laws with a range of examples, including the speed of sound in air, the shape of the Earth, the moon's pull on the Earth, and the orbits of the comets. All of this made him famous, especially the tale that he had come up with his theory of gravitation when an apple fell on his head, although that is now thought unlikely. It's more likely that he was musing on why an apple didn't go upwards or sideways when it fell from a tree.

Newton's great work was *Philosophiae Naturalis Principia Mathematica*, referred to simply as the *Principia*. Seen as the first great book of science and mathematics, it was hugely influential. It was published by the Royal Society in 1686, but there was a problem. The Society had spent its budget for books that year on a book called *De Historia Piscium* (*Of the History of Fish*) by one Francis Willughby. Edmond Halley, another astronomer and mathematician – and a big fan of Newton – saved the day and helped pay for the publication of Newton's book. Meanwhile *De Historia Piscium* was expensive to publish and sold poorly. Halley's salary from the Royal Society was cut, and he was paid instead with copies that hadn't sold. I'll bet Halley was over the moon. Nowadays you can pick up an original copy of the failed fish book for around $12,000. In 2016, a first edition of *Principia* sold for $3.7 million.

As it happens, it was Halley who made the next leap in astronomy. Using Newton's laws of motion, he figured out that a comet that appeared in 1682 had previously appeared in 1607, and also in 1531. By measuring the arc formed by the comet in the sky, he concluded that it was in orbit around the solar system and coming around every 76 years. He predicted it would return in 1758. He died in 1742 – but guess what, it did return in 1758, and it was the first object shown to orbit the sun rather than the planets. Proof that Kepler and Newton had got it right. This is one of the first examples of the predictive powers of Newton's laws of motion. Prophecy, but based on mathematics.

Halley commanded the first-ever purely scientific naval voyage in 1698, when he took charge of the *Paramour*, which was tasked with sailing to the South Atlantic to study the Earth's magnetic field. This can be seen as the first starship *Enterprise*. Unlike Captain Kirk, however, Halley's officers turned against him because of how he treated them, and he had to take the ship back to England. He is perhaps one of the most famous astronomers, because of the comet which bears his name, and also because of that most iconic of rock and roll bands, Bill Haley and His Comets. Mind you, I can't help thinking 'Shatner's comet' has a better ring to it . . .

What happened next was a lot more descriptions of planets, stars, and finally – as the power of telescopes increased – galaxies. What began as an instrument invented by Lippershey which could magnify a minuscule three times, got more and more powerful, allowing astronomers to see further and further into space.

In 1781, William Herschel discovered Uranus. This planet still makes students titter. He described over 2,500 celestial objects and worked with his sister Caroline, who was the first woman to discover a comet and became a minor celebrity in her own right. She was also the first paid female scientist, after being given a stipend from the Royal Astronomical Society in 1835. In 1838 she was made an honorary member of the Royal Irish Academy.

Jump!

At the beginning of this chapter, people 200,000 years ago were looking up at the sky and wondering what was going on up there. Records are thin, so we don't know the gender of those earliest astronomers, but it's safe to assume everyone was in on this. As we've seen, 'science' itself didn't come along until much later, but by then the patriarchy was entrenched and it was often men doing the science. But by the turn of the twentieth century, women were beginning to make major discoveries, although they often weren't given the credit.

Annie Jump Cannon was born in Delaware on 11 December 1863, and she would eventually be recognized as having discovered more stars in a period of four years than all the men had managed up to that point. She even came up with a whole system for classifying stars.

Cannon's mother had taught her about constellations and encouraged her to study science. At Wellesley College she studied under Sarah Frances Whiting, one of the few women physicists in the US at that time. She then studied at the Harvard Annex (now Radcliffe College), which was set up so that Harvard professors could repeat lectures to women students. In 1896 she became one of the Harvard Computers, a group of women hired by Harvard astronomer Edward C. Pickering to map and define every star in the sky. The men operated the telescopes and took the photographs, but the women examined the images and did the calculations. Cannon published her own work and began to be recognized for her brilliance. She became the first woman to receive an honorary doctorate from a European university, the University of Groningen. She classified around 350,000 stars. The American Astronomical Society now presents an annual award in her name to a distinguished female astronomer.

Henrietta Swan Leavitt was another 'computer', and in the early 1900s came up with a way to measure distances to faraway galaxies. She was paid 30 cents an hour. This isn't as bad as it sounds, as it was 5 cents an hour more than most of the other 80 women employed as computers and 50 per cent higher than the average wage at the time. By that stage astronomers had figured out that stars were very, very far away, using light years to measure the distances. A light year is the distance light travels in one year, which is 9.46 trillion kilometres. It was known that the distances to relatively close stars (up to a few hundred light years) could be calculated using the change in parallax (displacement compared to faraway stars) when viewed from two distant observatories on Earth. By examining sequential photographs of the night sky, Leavitt accurately measured distances to stars up to 20 million light years away. That is an unbelievably vast distance. The scale of the cosmos for us tiny humans is beyond comprehension, to me at least.

Bang!

Leavitt had a major influence on Edwin Hubble. Born in 1889, in his younger days he was a talented athlete, being especially good at basketball. Perhaps the spinning of basketballs on his fingertips is what inspired him to go into astronomy. He studied law at the University of Chicago, leading the university's basketball team to a national championship victory in 1907. He was also a powerful boxer and once knocked out the German heavyweight champion. He was one of the first to be awarded a Rhodes Scholarship, a highly prestigious postgraduate award for overseas students to come to Oxford University, and he took some science courses while there. On his return to the US, like the Great Gatsby, he seemed to reinvent himself, adopting an English accent, going around in a cape and with a pipe, and saying he had once been a great lawyer.

In 1917 he volunteered for the US Army and rose to the rank of major, although he never saw action. He took the opportunity while in Europe to study astronomy at the University of Cambridge. In 1919, he took up a post at the Mount Wilson Observatory in California, where he worked until his death in 1953. He did all his great work there.

When he arrived, Mount Wilson's enormous telescope had just been installed. At the time, it was thought that the universe just consisted of the Milky Way. How wrong they were. Using the work of Henrietta Leavitt, Hubble provided evidence that spiral nebula – so-called because they were in the shape of a spiral – were too far away to be in the Milky Way, and in fact were entire galaxies themselves. He was only 35 at the time, and the old guard initially refused to accept his findings.

Hubble's work on galaxies revealed that the universe was expanding, and this in turn provided evidence to support perhaps the greatest theory when it comes to the universe: the Big Bang theory. Hubble, along with Belgian astronomer Georges Lemaître, concluded that the further away galaxies are, the faster they are moving away from the Earth. If this expansion is extrapolated backwards, we get to what is termed a singularity, in which space and time lose meaning. We now know that, around 13.799 billion years ago, this singularity cooled sufficiently to allow subatomic particles to form, and then atoms. Gravity then kicked in, and atoms of hydrogen, helium and lithium clumped together – and hey presto, the first stars started to form. Hubble had come up with a start date for the universe.

Confusingly, the Big Bang is seen as not an explosion of matter moving outwards but instead space itself expanding, a bit like blowing up a balloon. For this to work, an as-yet-unknown 'dark matter' is needed to exert gravitational effects.

Dark matter is defined as matter that doesn't interact with the electromagnetic field, and so is difficult to detect. The evidence for dark matter is currently indirect, based on the observed movements of galaxies and the bending of light from distant galaxies. The Big Bang is effectively the beginning of physics and chemistry, which eventually led to biology.

Newton's work on gravity described how all objects exert a force that attracts other objects. His laws could predict the motion of planets and objects on Earth, and still do. But astronomers also observed that Newton's laws didn't work for some things, such as Mercury's odd orbit around the sun, as it moves faster than Newton's laws would predict. Enter Albert Einstein.

Einstein's Fudge Factor

In 1917, Einstein described how instead of exerting an attractive force on each other, objects actually curve the fabric of space and time around them, forming a dip in this fabric that other objects fall into. A reasonable analogy that's sometimes used is a bowling ball on a mattress. It will press into the mattress and draw objects towards it. In the same way, the sun draws planets towards it.

Einstein's prediction of the curvature of what became known as the space-time continuum was confirmed in the 1919 solar eclipse, when the path of light from a distant star was shifted by the sun's pull – by the exact amount that Einstein's law of general relativity predicted. What had been somewhat theoretical now had proof.

Albert Einstein was born in Ulm, Germany, and at the age of 12 had already taught himself algebra and geometry, discovering his own proof of Pythagoras's theorem. He remembered thinking at that age that nature could be understood as a 'mathematical structure'. At that age I had only managed to stain my bedroom ceiling brown . . . Einstein studied mathematics and physics at the Eidgenössische Technische Hochschule, the public university in Zürich (what a mouthful – it makes my own Scoil Bithchemic agus Imioneolaoicht, Colaiste na Trinoide seem inadequate) and got a job as a patent clerk at the Swiss Patent Office in Bern. While there, he worked on his theory of special relativity, publishing key papers in 1905. This theory describes the relationship between space and time, in the famous equation $E = mc^2$, where E is energy, m is mass, and c is the speed of light. He also published work that would influence the development of the field of quantum mechanics, which concerns the physics of atoms and subatomic particles and then involved such luminaries as Marie Skłodowska Curie and Erwin Schrödinger (more on them later).

By 1915 Einstein had discovered the theory of general relativity, and by 1929 there was such excitement about his talk being held at the American Museum of Natural History on the subject that extra police had to be called in – because a crowd of 4,500 'broke down iron gates and mauled each other'. If only science could be as exciting now.

It's almost impossible to explain Einstein's theory of relativity in lay terms. When a journalist asked him to explain in simple words time dilatation and Lorentzian contraction, he just said, 'They are technical terms.' Relativity is just too spooky. One clear experiment, though, illustrates the theory well. In 1971, Joseph Hafele, a physicist, and Richard Keating, an astronomer, proved that when an object moves, time slows down relative to an object that is moving more slowly. They took four atomic clocks – which are extremely accurate – aboard commercial airliners. They flew twice around the world, first eastward and then westward. This was felt to involve a sufficient length of time to determine if motion slowed down time. They compared the clocks against others that were on land, at the United States Naval Observatory. Time had slowed for the clocks that were moving fast. The time differences were tiny – if we fly in an airplane, time does slow down slightly for us, but it's too small a difference to be important. Time is therefore not moving forward in a constant fashion, but instead is relative. Though we still have no idea why that's the case.

Clearly Einstein's work had an influence on astronomy, particularly on the work of Hubble. In 1915 Einstein found that his theory of general relativity indicated that the universe must be either expanding or contracting. Quite a difference between those two, don't you think? He couldn't tell which, and so he introduced a 'fudge factor' into his equations. But once he learned of Hubble's work he realized that the universe must be expanding, and he saw it as the biggest blunder of his life. He'd even visited Hubble to discuss an expanding universe.

Despite this, general relativity went on to have as big an impact as Newton's laws, especially in the realm of astronomy. It led to the knowledge that galaxies can cluster into giant superclusters, and that within them all kinds of interesting structures exist, including neutron stars (which can pack all of a star's mass into something the size of a city; in more domestic terms, a sugar-cube-sized piece of neutron star would weigh as much as Mount Everest – heavy, man, but sweet) and black holes, which are so dense that not even light can escape their gravitational pull.

In 2019, scientists got an image of a black hole at the centre of galaxy M87. Katie Bouman was one of the scientists involved; she wrote a computer program to help create the image. The black hole was 500 million trillion kilometres from Earth. Bouman said she watched in disbelief as the first-ever image of a black hole appeared on her computer screen. The scientists observed the warping of space-time around the black hole, yet again proving that Einstein's theory was correct. Some black holes are billions of times the mass of the sun, and when they interact with a star they can release significant amounts of energy. They occur at the centre of most, if not all, galaxies.

When black holes collide, they send out what Einstein called 'gravitational waves' – although he only theorized about them, and it took until 2015 for the Laser Interferometer Gravitational-Wave Observatory (LIGO) to detect them. Black holes were very trendy things to work on because people had all kinds of ideas about them. What if you could travel through one of them? Would time slow down? Might you come out the other side in another part of the universe? Science fiction writers have had a field day with black holes.

Jocelyn Has the Last Laugh

Pulsars are another astronomical feature that has fascinated astronomers and science fiction writers. The first pulsar was discovered by Jocelyn Bell Burnell in 1967. She was from Lurgan, Northern Ireland, and as a child her father used to take her to Armagh Planetarium (which he had helped design). She went to Lurgan College, but like the other girls she wasn't allowed to study science. She was expected instead to study cooking and cross-stitching. But her parents protested, and cross-stitching's loss was astronomy's gain. She failed the eleven-plus, a key exam that allowed students to pursue an academic education, but was encouraged by a physics teacher. She became a PhD student in astronomy at Cambridge, first working on quasars, which are supermassive black holes surrounded by gases. They really are called 'supermassive', so they must be great. They had been discovered in the late 1950s as sources of radio waves.

Bell Burnell was studying radio waves one night at the Interplanetary Scintillation Array just outside Cambridge (now there's a place to work), when she detected what she called a 'bit of scuff' on her charts that recorded radio waves. She noticed that the bit of scuff was cropping up with great regularity – one pulse every one and a third seconds. This intrigued her, and she called it LGM-1, which stood for 'Little Green Man 1'. The source was eventually identified not to be an alien, but a rapidly rotating neutron star, sending out a pulse of radio waves like a lighthouse sends out a pulse of light as it rotates. The media took great interest in this finding, but Bell Burnell was dismayed to find that they mainly wanted to interview her PhD supervisor Antony Hewish, and only asked her about her hair colour and how many boyfriends she'd had. It was a *Daily Telegraph* journalist who called them pulsars, short for 'pulsating radio source'. Hewish went on to win the Nobel Prize in Physics, although that didn't bother Bell Burnell, who said, 'I am not myself upset about it – after all, I am in good company, am I not?' She had the last laugh, though, winning the Special Breakthrough Prize in Fundamental Physics in 2018, which came with prize money of $3 million. She donated all of this money to support women, under-represented ethnic minorities and refugee students in physics.

You're Never Alone in the Multiverse

You will have noticed that we are getting pretty close to today, and the latest exciting discovery in astronomy is exoplanets – planets outside our own solar system. The first definite one to be discovered was in 1992, but over 5,000 of them have now been found in over 3,500 planetary systems, many of which have more than one planet. About one in five sun-like stars have Earth-like planets in the habitable zone, meaning the conditions are reasonable for life to evolve, including the likelihood of there being surface water – but none appear hospitable for life so far. Currently, there are estimated to be 5 billion possibly habitable Earth-like planets in the Milky Way alone. Exoplanet hunters have even listed 29 exoplanets where an alien observer could have discovered Earth in the past 5,000 years, by watching it transit across the face of the sun. They are close enough to intercept radio or TV broadcasts that began 100 years ago.

The discovery of exoplanets has re-energized the search for extraterrestrial life, and many astronomers believe that life must exist out there somewhere, because there are likely to be billions and billions of exoplanets, all rolling the dice of biochemistry, which eventually might give rise to a living cell. The next question is whether that cell might evolve into so-called intelligent life. The Search for Extra-Terrestrial Intelligence (SETI) Institute has since 1984 been scouring the sky for signs of intelligent life. Over 100 scientists are doing all kinds of experiments and analyses, including trying to find sources of electromagnetic radiation, which might be emitted by aliens out there somewhere and would be evidence of advanced civilization. If they ever find life out there, and they are confident that they will, I suspect it will make us feel just that little bit less special.

In September 2015, the Hubble Space Telescope, which is in orbit around the Earth, detected the most distant galaxy found to date. Hubble can observe in the ultraviolet, visible and near-infrared parts of the spectrum. As Hubble is in orbit around the Earth it has much greater sensitivity than Earth-based telescopes, because of less light pollution. The most distant galaxy we have observed is 13.5 billion light years away – and remember, a light year is 9.46 trillion kilometres. But the current view is that the universe has a diameter of 93 billion light years, with 90 per cent of the galaxies yet to be discovered. The one we're in, the Milky Way, is part of the Laniakea Supercluster, which alone has 100,000 galaxies. *Laniakea* is a Hawaiian word meaning 'immense heaven'. That just about sums it up alright. All of those galaxies have billions upon billions of stars, and many of those will have planets orbiting them.

Aristotle thought that the entire universe was within eyesight, and therefore quite small. And why wouldn't he? He didn't have a telescope. This all began when Lippershey made a device that could bring things far away a bit closer. But only three times closer. The telescopes we now have bring things billions of times closer – and in fact, when the Hubble telescope looks up it is looking into the past, because of the time it takes light from those distant galaxies to reach it. Hubble can even see 'toddler' galaxies that only formed around 1 billion years after the Big Bang. Nice work, Hubble. But remember how impressed you were by your iPhone 11 and its camera, until the iPhone 13 came along? Well, the James Webb Space Telescope launched in 2021, and all the cool space agencies don't want to be seen with a Hubble anymore. If the Hubble can see toddler galaxies, then the James Webb can see baby galaxies – born within minutes of the Big Bang. The mission of the telescope is to study the formation and evolution of galaxies and stars, to search for light from the first stars around 100 million years after the Big Bang and provide information on the possible origin of life. In other words, to get to the core of where we came from and perhaps where we're going. A stunning objective, but not as stunning as the first images from Webb, which were as spectacular in their clarity and beauty as they were in their data content.

And astronomers are even going beyond our own universe, exploring the multiverse, a still-hypothetical group of multiple universes existing outside our own Big Bang. These different universes are called parallel universes. In 1952 in Dublin, Erwin Schrödinger gave a lecture which he warned might 'seem lunatic'. He effectively said his equations suggested that lots of universes might be happening simultaneously. And evidence has been growing, with lots of astronomers and mathematicians getting involved. Cosmologist Laura Mersini-Houghton is a proponent of the theory of multiverses, with evidence coming from a 'cold spot' called the Eridanus supervoid, in the cosmic microwave background (or CMB, if you want to show off), which is the radiation that permeates the observable universe thought to be a remnant of the original Big Bang. A region of slightly lower temperature and lower density of matter and dark matter in the CMB might have been created when the different universes separated.

Mersini-Houghton sees the multiverse as a natural

extension of what Copernicus described: we once thought the Earth was the centre of the universe and it now turns out that even our own universe isn't central, it's just what she describes as 'one tiny grain of dust in a much more intricate and beautiful cosmos'. Scientifically, the idea of the multiverse speaks as ever to the power of the individual scientist working in the far reaches of human knowledge, coming up with bizarre concepts that might explain observable phenomena in a way that can still be proven scientifically, no matter how weird they seem.

What would those early astronomers who built Newgrange more than 5,000 years ago make of all this? Every time astronomy has made an advance – beginning in earnest with Galileo – we have become more insignificant. No wonder the Catholic Church condemned him. We went from the Earth being the centre of the universe, to being just another planet going around a star – the sun. Now we know that our star isn't special either; it's just one of countless billions of other stars, in billions of galaxies, many of them with planets like Earth going around them. Life itself will most likely not be unique in the universe, which itself may not be unique, but rather one of countless universes.

Perhaps there's an astronomer out there on a habitable exoplanet looking at the Earth and wondering if life exists here. There may even be someone out there writing a chapter on astronomy in a book just like this one – and if so, I hope it sells better than Francis Willughby's history of fish.

2. Build a Rocket, Boys and Girls

All that astronomy might seem to be a lot of navel (or star) gazing – and much early astronomy was performed at sea, so it was also naval gazing – but it was actually useful in all kinds of ways, not least as the starting point for all science. But while looking at the sky was all well and good, the real fun thing to do would be to go and visit. We humans love to travel. Our history tells us we are a nomadic species, spanning out from Africa and moving all over the world. The sky and beyond was the next frontier, and hopefully we would come back safe and well to tell the tale.

Flying Sheep

In June 1783, two brothers, Joseph-Michel and Jacques-Étienne Montgolfier, sent a large balloon up into the Paris sky. This is a great example of how you can be good at one thing – they were paper manufacturers – but then have dreams well beyond that, in this case to fly using a craft . . . made of paper. Balloons are key to parties, as I well remember from blowing up hundreds over the years for my sons' birthdays. The stress of it. Will it burst? Can I tie the damn thing off? But the brothers Montgolfier had bigger ambitions for their balloons – they wanted to fly. And so the space race began with unmanned flight, followed by flying animals – they sent up a duck, a rooster, and a sheep called Montauciel (which apparently means 'climb to the sky'). Imagine the mayhem of trying to get a duck, a rooster and a sheep into a balloon.

This flight was performed at Versailles in front of King Louis XVI, and his wife Queen Marie Antoinette. It must have been some spectacle. The brothers brought in Jean-Baptiste Révillon to help. He was a wallpaper manufacturer and he decorated the balloon with golden flourishes, zodiac signs and suns. Dressed in their finery, the king and queen and other members of the aristocracy – plus 130,000 spectators – turned out to watch this big brightly coloured balloon.

The sheep was thought a good approximation for the weight of a human. The duck was included as a control, as it could already fly up in the sky, and so any effects that might occur on it could be put down to the balloon. The rooster was another control, since it was a bird that couldn't fly.

Controls such as these are extremely important in science as they allow us to interpret results correctly. In every laboratory experiment, including in my own lab, we include what are called 'negative' and 'positive' controls. Let's say we're trying a new medicine to see if it will block a disease-causing process in the body. The negative control will leave the medicine out of one set of tests, and so no effect should be observed. Meanwhile the positive control is a medicine that should work, perhaps targeting a different part of the process than the one I'm trying to block, but which will have a similar inhibitory outcome. If the negative control has an effect, there's something wrong – any effect seen is 'non-specific', as any perturbation of the system might stop the disease process in question. Whereas if the positive control fails, that means there's something up with the experimental system, as there is not the expected inhibitory effect. The experiment therefore wasn't set up in the correct way.

Applying this to the creatures in the Montgolfier experiment, if the sheep got sick but the rooster or duck didn't, that might mean humans would too, because sheep are mammals like us. If the sheep and the rooster got sick, then any creature that can't fly, including humans, might be harmed. In the end none of the animals got sick in this experiment, so all was well.

And so, possibly watched by a relieved sheep, duck and rooster, on 19 October 1783 two of their friends climbed into a balloon still tethered to the ground, lit a fire under the neck of the balloon and took off: the first manned flight. This flight took place above the Folie Titon in Paris, a city obviously way ahead in the Balloon Race.

On 21 November 1783, the first clearly recorded untethered manned flight took place. Louis XVI got more involved this time and insisted that convicted criminals should be the first pilots, to the annoyance of the Montgolfier brothers, who wanted to be the pilots themselves. Criminals, however, were seen as dispensable should anything go wrong. The balloon travelled for 25 minutes and flew over seven kilometres.

And then, just a few days later, the first Montgolfier flights were emulated by Jacques Charles and co-pilot Nicolas-Louis Robert over the Jardin des Tuileries. They got to a record height of 550 metres and landed at sunset, 36 kilometres away. They carried a thermometer and barometer to measure the temperature and pressure of the air, providing the first meteorological analysis of the atmosphere well above the Earth's surface.

Charles decided to ascend again on his own. The sun was

starting to set, and he rose to 3,000 metres until he could see the sun again. This was evidence that the Earth indeed was round, which was still disputed by some at that time (and still is today, as we'll see later). Four hundred thousand people witnessed this launch. There must have been a lot of oohing and aahing.

Jacques formulated what would later become known as Charles's law, to explain the relationship between the temperature and volume of a gas – showing that hot air occupies a larger volume than cold. Thus a balloon filled with hot air has a lower density than the air outside the balloon, resulting in it floating upwards. Hydrogen gas has a lower density than the main gases in air, nitrogen and oxygen, and so does not require heating to make the balloon buoyant. Hydrogen gas for a new Charles/Robert balloon was made by pouring almost a quarter of a ton of sulphuric acid onto half a ton of scrap iron. In chemistry terms, Fe (iron) reacts with H_2SO_4 (sulphuric acid) to make $FeSO_4$ (iron sulphate) and H_2 (hydrogen gas).

Daily press bulletins were issued, and many people turned up to watch all of this, as it must have seemed like a mad pursuit. All that bubbling and the capturing of the gas. And then, on 27 August 1783, the first unmanned hydrogen balloon was released. Benjamin Franklin, the American scientist and statesman (more on him later), was on hand to watch as the balloon lifted off the ground and flew north for 45 minutes, pursued by people on horseback. Charge! It landed 21 kilometres away, where terrified locals attacked it with pitchforks. Charles and Robert went up again, this time in the first manned hydrogen-filled balloon.

Balloon mania exploded, with lots of balloons being made and flown. This led to the world's first aircraft disaster, which happened in Ireland in May 1785, in Tullamore, County Offaly. A hot-air balloon crashed, leading to 100 houses burning down. A newspaper report from the time blamed 'an English Adventurer' for the accident. The town still has the phoenix rising from the ashes as its emblem.

Charles Green was a notable balloonist from London. In 1828, he claimed that he had taken his horse up with him in a balloon, but this was disputed. Why he might have wanted to take his horse up in a balloon was not recorded. A whole 22 years later, in 1850, he did finally manage to get a pony into a balloon with him. The business of bringing an animal with you became something of a craze, and in 1852 one Madame Poitevin took a bull with her. She dressed the bull up as the god Zeus. She said she wanted to be like the Greek goddess Europa, who rode on the back of a bull. It seems as if she was trying to outdo her husband, who was another one who'd gone up in a balloon with a horse. Quite what she did with the bull when she was up in the balloon is not known, but a local newspaper commented that a mob gathered and were 'astonished to see a bull fall from a balloon – a thing not seen every day in Essex'. She was charged with cruelty to animals and fined five pounds. The lifting of animals in balloons stopped.

It didn't stop the Poitevins though, who a month later went up together in a balloon reaching 5,000 feet, at which point Monsieur Poitevin jumped out of the balloon with a parachute. Whether that was to get away from his wife wasn't recorded, but he survived. We take this sort of thing for granted today, but falling out of the sky and not dying must have been amazing to witness in the 1850s.

Ballooning allowed humans to rise above the Earth and look down, rising to altitudes that even birds had trouble reaching. But this wasn't enough. The aim moved to going higher for longer and with more control.

Bluffeurs!

Orville and Wilbur Wright invented and flew the first airplane, in Kitty Hawk, North Carolina, in 1903. Their interest in flying had begun when their father gave them a toy made of paper, bamboo and cork that had a rubber band for its rotor, and which flew like a helicopter. The two brothers played with it until it broke and then made their own.

While still a teenager, Wilbur was struck by a hockey stick which knocked out all his front teeth. He didn't like going out after that, and he spent a lot of time reading in his father's library. His brother Orville, on the other hand, dropped out of high school and set up a printing shop. Wilbur joined him in the business and the brothers launched a newspaper called *West Side News* in their home town of Dayton, Ohio.

The 1890s saw the popularization of the 'safety bicycle', with a chain-drive transmission and gears. Powered by the legs via pedals, the links in a flexible chain connect with the teeth on one circular chain ring and transfer energy to a second chain ring linked to the axle of a wheel or propeller. The relative circumference of the two rings, or gear ratio, regulates the speed of the axle. This invention led to a bicycle craze in the US in which millions of people bought bicycles, and the bike boom was especially significant for women. Cycling increased women's mobility outside the home, which among other things allowed them to meet (and marry) a broader range of people than those found in their home towns, transforming gender relations. Many women bought bicycles, and this gave rise to the 'rational clothing' movement, with less restrictive skirts. A special 'bicycle wear corset' was even invented. This change in women's clothing might seem minor today but it was radical at the time.

The Wright brothers were well aware of the demand, and decided to open a bicycle sales and repair shop, as well as making their own bicycles. They also kept publishing newspapers and magazines, and were inspired by articles about Otto Lilienthal, a German who had made gliders. In August 1896, Lilienthal was killed when one of his gliders crashed to the ground. His death focused the brothers' minds on a key issue: how to control the wings of a glider. Watching birds, Wilbur figured out that birds change the angle of the ends of their wings to move left or right.

And so they began to design an aircraft. They made the wings of the aircraft they designed movable, and in 1900 they went to Kitty Hawk, North Carolina, to start their manned gliding experiments. They chose the location because of regular breezes and soft sandy surfaces to land on. A key feature was to put a 'camber' into the wings – a curvature of the top surface. Lift was achieved by the air flowing more quickly over the upper side of the cambered wing than the underside. The brothers' first flights with the glider were unmanned, but then Wilbur climbed on board and was able to fly the glider over distances of a few metres.

In 1901 they returned to Kitty Hawk and built gliders with larger wing areas. These proved difficult to control, and on their way home from that year's efforts, Wilbur said to Orville that man would 'not fly in a thousand years'. As with probably all inventors and indeed scientists, they had reached a low point and almost quit. But instead they went back to the drawing board. By 1903 they were ready to test their new design, with an engine attached to drive a propeller (which had been used in boats) to achieve lift-off. The propeller was used to achieve 'thrust' – the pulling of air behind the propeller to create a pressure difference, pushing the aircraft forward – which they knew was needed to counteract the drag caused by gravity. And it worked!

The aircraft was called the *Wright Flyer*. Wilbur won a coin toss to see who would go first – and his attempt lasted three seconds. On 14 December, which by coincidence was the 121st anniversary of the Montgolfier brothers' first unmanned hot-air-balloon flight, the aircraft did better; and on 17 December, Orville flew 120 feet in 12 seconds. They then took turns, with one flight lasting 59 seconds over 852 feet. The *Wright Flyer* was the first airplane – a machine that could fly over the ground under the control of a pilot.

The brothers sent a telegram to their father asking him to 'inform the press'. He contacted the *Dayton Journal*, but they refused to cover the story, saying that the flights were too short to be important. A telegraph operator leaked the news to a Virginia newspaper – suitably called the *Virginian-Pilot* – and the story was reprinted in several newspapers, including the *Dayton Journal*. The problem was the article was a major exaggeration. The journalist had clearly felt the news had to be 'sexed up', and the headline read: 'Flying machine soars three miles in teeth of high wind'. It did say 'No balloon attached', which was true. The Wright brothers issued a statement saying the article was 'fictitious' and gave their own account, which few newspapers published, deeming it less interesting.

This is not unlike my own experience with the media. When my lab makes an important discovery, we usually issue a press release to inform the public, who are after all usually funding the work from their taxes via the Irish government. Most of the time the coverage is accurate. But once I was interviewed about a finding my lab had made about an immune-system protein called Mal. When bacteria are sensed by immune cells like macrophages, Mal activates the cells to help them kill the bacteria. I had won an award for the discovery, and when I explained it to a journalist, I said that in some ways I was like a car mechanic trying to figure out how a car engine worked. The headline of the newspaper article was 'Car mechanic wins science award'.

The Wright brothers kept going. On 20 September 1904, Wilbur flew in a circle, and by the end of the year they had flown for a total of 50 minutes. Journalists had lost interest, with one of the few accounts being published in a beekeeping magazine of all places. One well-known journalist wrote, 'Frankly, none of us believed it.' The prominent magazine *Scientific American* questioned the 'alleged experiments'. And an article in a French newspaper called them *bluffeurs* (bluffers). The Wright brothers didn't mind, though, because by that stage they were fearful of competitors. Negative press suited them, as it spurred them on to make progressive improvements.

But interest grew and aircraft design continued to advance. On 14 June 1919, some fifteen years after the Wright brothers' success, John Alcock and Arthur Brown were the first to fly across the Atlantic, from St John's, Newfoundland to Clifden, Ireland. In Dublin they were carried over the heads of a group of students and taken to the Dining Hall of Trinity College Dublin for a feast. It took Alcock and Brown 16 hours and 28 minutes to cross the Atlantic, a trip that took 15 days by steamship. Then, on 9 June 1928, Charles Kingsford Smith landed in Brisbane, having flown across the Pacific in three stages. He was also the first to circumnavigate the world, crossing the equator twice. These were heady days. Suddenly the world became much smaller.

Over the two and a half decades since the first flight, the original piston-engine-driven propeller technology of the Wright brothers had steadily been improved, achieving more thrust and therefore aircraft speed. But in the 1930s engineers realized that engines driving propellers were reaching a limit (around 460 miles per hour – imposed by mechanical stresses on the propeller tips as speed increased). In 1929, Frank Whittle submitted his ideas for a jet engine to the Royal Air Force in the UK. It involved something called a gas turbine, but the RAF weren't interested. German engineers had similar ideas, and at the end of the Second World War the Russians and Americans used some of the designs to make jet engines. These work by sucking air into the front of the engine using a fan. The engine then compresses the air, mixes it with fuel, ignites it and shoots it out the back, thereby creating thrust.

Airplanes could now go further, and at greater speeds. Later, engines that allowed planes to fly faster than the speed of sound were developed. Due to the amount of fuel required, they proved too costly to operate for routine civilian use; the ill-fated Concorde ceased flying in 2003.

Arrows of Flying Fire

Airplanes were going further and faster, but they couldn't take anyone to space. That required an even more powerful machine: a rocket. The use of rockets goes way back, to before the jet age. In 1232, the Mongols and Chinese were at war, and during one battle the Chinese fired what were described as 'arrows of flying fire'. These rockets comprised a tube containing gunpowder, which was set off by a long stick, releasing gases that propelled the rocket forward. The

Chinese often used them in their firework displays. They never thought of using them to fly into space, as far as we know, perhaps because they thought them too dangerous – or perhaps because it was over 700 years before the space race.

This next bit is literally rocket science. A Russian, Konstantin Tsiolkovsky, started the process that would give rise to the space age. In 1903 he published the rocket equation, which relates rocket speed to mass, based on how fast gas is emitted. Tsiolkovsky had been suspended from school at the age of 14, having received precious little education. This didn't stop him, however, and he self-taught himself mathematics, physics and chemistry. He was a fan of Jules Verne, whose novel *From the Earth to the Moon* inspired his interest in space flight. He built his own centrifuge – a machine that spins things at speed, creating a gravitational pull – which he used to study the effect of gravity on chickens. As with Montgolfier's animals, the names of his test subjects are unrecorded.

Tsiolkovsky began writing science fiction, but he found himself drawn to the prospect of space travel. In 1892 he built Russia's first wind tunnel, and in 1903, around the same time as the Wright brothers were tentatively taking off in their flying machine, he wrote an article titled 'Exploration of the World of Space with Reaction Machines', in which he set out how to explore space with rockets. In 1911, he proposed liquid hydrogen and liquid oxygen as a fuel source, which was actually used on the space shuttle decades later, in the 1980s.

His day job as a teacher distracted him from his research, and his next important work didn't appear until 1929, when he wrote about a multi-stage rocket – as each rocket uses up its fuel supply, it breaks off, reducing the mass of the remaining rocket. He also predicted the need for pressurized suits for astronauts.

Tsiolkovsky's writings proved useful in the effort to build rockets from the 1950s on, and the biggest crater on the far side of the moon is named after him. He had predicted humans would go into space, writing: 'Mankind will not forever remain on Earth, but in the pursuit of light and space will first timidly emerge from the bounds of the atmosphere, and then advance until he has conquered the whole of circumsolar space.'

While Tsiolkovsky was dreaming and writing about rockets and space, the American Robert Goddard began building them. In 1926 he built the first liquid-fuelled rocket, and held a patent for a three-stage rocket. Similar to Tsiolkovsky, he was inspired by science fiction, having read H. G. Wells's *The War of the Worlds*, about an invasion from Mars. It is vitally important for scientists to dream, and these scientists' dreams came from the science fiction books they had enjoyed in their youth.

Goddard studied at the Worcester Polytechnic Institute, where his most notable achievement was firing a rocket from the basement of the physics building, which he did for an important scientific reason: fun. This was fuelled by powder, like a Chinese firework. He later published a paper on rocket propulsion, writing that by his calculations a rocket could one day get to the moon, and suggested it should explode on the surface to show it had made it. *The New York Times* scoffed at his ideas, but after the 1969 moon landing the paper published an apology. Not unlike the apology that came from the Catholic Church to Galileo, but much quicker.

Goddard moved to Roswell, New Mexico in 1930, and launched 56 rockets over the 12 years he was there, providing useful information for future developments. It's unlikely his move to Roswell had anything to do with the aliens that are allegedly hidden there. Those rumours began in 1947 with the so-called Roswell incident, when officers from the Roswell Army Air Field recovered metallic and rubber debris. In the 1970s, one of the officers involved, Jesse Marcel, said that the debris was from an extraterrestrial spacecraft. No evidence for aliens in Roswell has ever been presented but that hasn't stopped the town from capitalizing on the conspiracy theories – the city's official seal has a little green man. There was a lot of secrecy around what the debris might be from, and this was finally explained when the US Air Force published a report stating that the crashed object was a nuclear-test surveillance balloon, the test having been carried out in secret. Problem solved. Oh, if only it were that easy . . .

Vengeance Weapons

In 1944, Wernher von Braun developed the V-2 rocket for the Nazis. This was a long-range rocket with a bomb on board, designed as a 'vengeance weapon' to attack Allied cities in retaliation for the bombing of Germany. These rockets were fired at Britain during the Second World War, killing thousands of people. If any more evidence were needed that the Nazis were bad guys, they deployed slave labour to build the V-2 rockets. More people died making them than were killed by them – around 9,000 civilians died in the bombings, but over 12,000 labourers died during production. Von Braun visited the factory and said it was 'repulsive'. He was arrested in 1944 when a young female dentist who was an SS spy reported that she had heard him say he would rather work on space rockets, and that the war wasn't going well. He was released because Albert Speer, then Minister of Armaments and War Production, intervened with Hitler, saying that von Braun was essential to the V-2 rocket programme.

At the end of the war, the Russian, British and US militaries gathered information on the German rocket programme and tried to recruit trained personnel. The US captured a large number of German rocket scientists including von Braun, and brought them to the US in a campaign known as Operation Paperclip. Over 1,600 German scientists and engineers were brought to the US.

In early 1945, with the Soviet Army closing in, von Braun had assembled his staff and said they should surrender to the Americans – he felt that the Americans would look after them better. The SS however caught up with him and 500 colleagues, and they were moved to the Bavarian Alps. Their SS guards were told to execute anyone trying to escape. It was only a matter of time before the Americans advanced to the new location, however, and so eventually von Braun and many of his colleagues surrendered to the Americans, stating: 'We knew that we had created a new means of warfare, and the question as to what nation we were willing to entrust this brainchild of ours was a moral decision more than anything else.'

Once in captivity he was interrogated – initially by a British engineer, L. S. Snell, who would go on to invent Concorde's engines. He was then transported to the US and was sent to an army installation near El Paso, Texas. Their presence was revealed in a magazine article headlined 'German scientist says American cooking tasteless, dislikes rubberized chicken'. While in El Paso, he refurbished and launched V-2 rockets that had been captured in Germany.

In 1950, von Braun and his team were moved to Huntsville,

Alabama, where he would stay for 20 years. He helped develop the first nuclear ballistic missile and witnessed its launch and detonation. He and his team also constructed the rocket Jupiter-C, which launched the West's first satellite, Explorer 1, on 31 January 1958.

But the Russians were making progress too. Sergei Korolev had trained as an engineer, as did many young Russians at that time, following Stalin's instruction that one in three students should be studying engineering. He began working on aircraft design for the Soviet military but in 1930 switched to working on liquid-fuelled rockets.

He was arrested in 1938, charged with slowing down the effort to develop rockets – although there was no evidence for that – and for being an intellectual during the Great Purge. He was sent to a gulag in Siberia where he lost his teeth because of scurvy, but was moved to a prison specifically for intellectuals and the educated, where he was given projects to complete by the Communist Party. In 1945 he was brought to Germany to recover technology used in the von Braun-designed V-2 rocket. Stalin made rocket and missile development a national priority in 1946 given their potential as weapons, and Korolev was promoted, overseeing a team of at least 170 German scientists who had been brought to Russia to continue their work on rockets, in a facility that was 200 kilometres from Moscow and surrounded by barbed wire, guards and dogs in case any of the Germans tried to escape. Korolev was made chief designer of long-range missiles. He initially made replicas of the V-2 rocket and began improving on it, making a rocket with a longer range which could reach England from Russia. The Germans, who proved difficult to work with, were sent back to Germany in 1951.

In July of that same year, two dogs – Tsygan and

Dezik – became the first space dogs, flying 110 kilometres straight up and down in an experimental rocket and living to wag their tails. Korolev then invented the world's first intercontinental ballistic missile, the R-7 Semyorka. It was a two-stage rocket and was armed with a nuclear warhead that could travel 7,000 kilometres. The nuclear arms race had begun. In 1957 Korolev was deemed 'fully rehabilitated' and the government apologized for arresting and imprisoning him.

Korolev began to plan for a rocket to launch a satellite. Eisenhower's administration got wind of this, probably from a spy, and in 1955 the US embarked on its own satellite programme. Korolev contacted the Soviet leadership asking for more resources, citing international prestige as a key motivation. Funding was provided and so began the Sputnik programme. This satellite was designed and built under Korolev's leadership in less than a month. It was a polished metal sphere around the size of a beach ball, carrying a transmitter and four external communication antennae. It was launched into space on a rocket on 4 October 1957. The scientists and engineers were relieved to pick up its 'beep-beep' radio signal once it was in orbit and called the Soviet president, Khrushchev, holding up the radio receiver for him to listen.

The Russians had issued information on the launch, and in the US 150 stations were on standby to detect the radio signal. Anyone with a shortwave radio could pick it up. The US government also started Operation Moonwatch, one of the first 'citizen science' projects. Amateurs were enlisted and equipped with small telescopes, and asked to time and track Sputnik's passing-over. The American Radio Relay League told enthusiasts to tune into 20 megacycles, saying that the 'beep-beep' sound would be heard each time it passed overhead. Engineers working for the company RCA made a recording and rushed to Manhattan to broadcast it over NBC radio. In the UK, the media reacted with a mixture of fear for the future (although of what wasn't especially clear) and amazement at human progress. A device had been launched into space which could orbit the Earth and send out a radio signal. Such excitement. Sputnik orbited for three weeks and then fell back into the atmosphere, burning up on re-entry.

The Soviets used the success of Sputnik to claim technological superiority over the West. American propaganda had been claiming that the US was a technological superpower, with the Soviet Union described as a backward place. Sputnik put paid to that and galvanized the US to work aggressively on rockets and space travel. The US launched its own satellite, Explorer 1, just over three months later. The US also set up the National Aeronautics and Space Administration, or NASA. Congress provided low-interest loans for college tuition for courses in maths and science, to encourage interest. Many who would join NASA were directly inspired by Sputnik – including Alan Shepard, the first American in space.

Khrushchev had become bored with another 'Korolev rocket launch', but when he saw the international reaction he became very interested indeed and began pouring resources into the rocket programme. The launch of Sputnik had stunned the world. Korolev excited Khrushchev further by proposing that they launch a dog into space. Sputnik 2 was quickly made and was six times bigger than Sputnik 1. It was successfully launched on 3 November 1957, with possibly the most famous dog ever as its passenger. The launch crew gave her the name Laika, which means 'Barker'. Presumably, she barked a lot. The US press called her Muttnik.

Laika was a stray mongrel who had been lured off the streets of Moscow by the rocket scientists with sausages. A dog of the people. She was the first animal to be put into orbit. Sputnik 2 had a life-support system which was an oxygen generator and a CO_2 absorber, and Laika was fitted with an electrocardiogram to monitor her heart rate. During the mission, at peak acceleration her breathing increased fourfold, and her heart rate doubled during the launch. There was no possibility of bringing Laika back to Earth, and the dog died from heat exhaustion after five hours in space, shortly before she was due to be poisoned. Poor Laika. One of the scientists had taken her home the night before the launch to play with his children, saying, 'I wanted to do something nice for her.' The technician who placed her in Sputnik 2 had kissed her nose and wished her bon voyage, knowing she wouldn't be returning. Five months later, after 2,570 orbits, Sputnik 2 – with Laika's remains on board – broke up on re-entry. Scientists didn't know if humans could survive a rocket launch, but as had happened with the sheep in the Montgolfier brothers' balloon, Laika had shown that they most likely could – at least for a while.

Lift-Off

At this stage of the space race, it was two-nil to Russia. They had launched a satellite and put an animal into orbit. And then they scored another major success. In 1958 Korolev began working on a spacecraft to carry a human into space. Vostok was designed to hold a person in a space suit. On 15 May 1960 a prototype was launched and orbited the Earth 64 times. Two dogs were launched next – Chaika and Lisichka, but sadly an explosion killed the dogs. Two other dogs, Belka and Strelka, were then launched into orbit, and even more impressively they were brought back down to Earth.

And then, on 12 April 1961, Yuri Gagarin became the first human in space. Korolev was the mission controller, speaking to Gagarin inside the capsule. He was the first cosmonaut, (from the Greek *kosmos*, 'universe', and *nautes*, 'sailor'). The Russians wanted a different word from the Americans, who were using 'astronaut', from the Greek for 'star'. Vostok 1, with Gagarin on board, completed one orbit of the Earth.

As a young man, Gagarin had worked in a steel plant, but then joined the Soviet air force. He had been fascinated by airplanes as a child, especially after seeing a Russian fighter plane crash near his village during the war. He was quickly spotted as a talented pilot and began flying the MiG jets produced by two engineers named Artem Mikoyan and Mikhail Gurevich (hence MiG). In 1959 he was selected for the Vostok programme. Twelve potential cosmonauts were selected, and when they were asked to vote for a candidate besides themselves, he was the most favoured. They were put through a range of physical and mental tests, including being put into an isolation chamber for hours on end and subjected to G-forces in a centrifuge. When the body is subjected to an increase in G-force, the heart rate goes up to keep the blood flowing, especially to the brain.

Gagarin emerged as the best candidate, and on 12 April at 6.07 a.m. he was launched into space. During the launch there was an exchange between Korolev and Gagarin.

Korolev: 'Preliminary stage . . . intermediate . . . main . . . LIFT OFF! We wish you a good flight. Everything is all right.'

Gagarin: 'Off we go! Goodbye, until we meet soon, dear friends.'

Poyekhali is Russian for 'off we go!' and became a popular expression in Russia, marking the start of the space age. Gagarin was promoted to the rank of major during the flight. Vostok 1 was in orbit for 108 minutes and then returned to Earth, over Kazakhstan. When it was 7,000 metres above the Earth, Gagarin ejected from the capsule and parachuted down. He became a national hero and worldwide celebrity, and the answer to a question in many a pub quiz. Millions of people turned up to see him at a special parade in Red Square. He visited 30 countries, but John F. Kennedy barred him from the United States, presumably because he didn't want him to be seen as a hero.

Gagarin went on to have a colourful life. He worked on the Soyuz space programme (the Soviet equivalent of NASA) but was permanently banned from training pilots after Soyuz 1 crashed. In 1961, following his rise to fame, his wife caught him in a liaison with a nurse who had helped him after a boating accident. He tried to escape, jumping from a second-floor window, resulting in an injury which left a permanent scar over his left eyebrow. He died in a plane crash in 1968. Three-nil to the Russians. But NASA would win the next big prize: putting a human on the moon.

Whoopie Man!

Because of Sputnik, NASA had been given a budget of $100 million, the equivalent of $1 billion today. The mission to get an American into space was named Project Mercury and candidates were selected from the US Navy, Air Force and the Marines. On 5 May 1961, Alan Shepard became the first American in space, less than a month after Gagarin's success.

Alan was obviously the right stuff. He could trace his ancestry back to the *Mayflower* and had served in a naval destroyer during the Second World War. One of his first assignments was to observe the launch of the rocket SM-65D, which was similar to the one that would bring the lucky person chosen into orbit. After lift-off it exploded, leading Shepard to remark to his fellow astronaut John Glenn (who would go on to be the first American to orbit the Earth): 'Well, I'm glad they got that out of the way.'

But his wife, Louise, had an even better sense of humour. When he was chosen, Alan told her that she had her arms around the man who would be the first man in space. She said: 'Who let a Russian in here?'

When Shepard heard that Gagarin had made it, he slammed his fist on the table so hard that people thought he had broken his hand. On 5 May he finally made it, in a spacecraft he named Freedom 7. The launch was delayed somewhat, and Shepard had to relieve himself in his spacesuit. He was uncomfortably wet for most of the mission. This led to future spacesuits having a 'waste collection feature'.

Shepard was in space for 15 minutes, but unlike Gagarin he didn't go into orbit. However, he had control over his spacecraft, whereas Gagarin's had been automatic. Shepard's waggishness was further in evidence when he answered a reporter who asked him what was going through his mind just before lift-off. He replied: 'The fact that every part of this ship was built by the lowest bidder.'

Shepard was given a ticker-tape parade on his return. They mustn't have worried about litter too much in those days. The Mercury mission was deemed a success and so was shut down, despite Shepard making a personal appeal to John F. Kennedy. He would subsequently lead the Apollo 14 mission, the third to land on the moon. In February 1971, he landed the Antares lunar module on the surface of the moon. On that mission he achieved another first: he hit two golf balls on the moon using a six-iron head attached to a lunar sample scoop. He said he drove the second ball for 'miles and miles and miles'.

Before that, of course, there was the small feat of Neil Armstrong becoming the first man to walk on the moon, on 20 July 1969. His flying career started in the US Navy and he saw action in the Korean War, flying 78 bombing missions and receiving several medals for his service. He then became a test pilot, flying over 200 types of aircraft, and joined NASA when it started, having been selected for the Man in Space Soonest programme. He applied to join the Gemini project in 1962 but missed the deadline. Luckily, a friend slipped his application into the pile without anyone noticing. One of Gemini's goals was to practise space rendezvous and prepare for a seven-day mission, which would be required for a moon shot.

President Kennedy had made a now-famous speech at a special joint session of Congress in 1961, saying that there would be an American on the moon before the end of the decade. Clearly, he didn't want to be four-nil down to the Russians, who were also in the race. In a follow-up speech Kennedy said: 'We set sail on this new sea because there is new knowledge to be gained . . . We choose to go to the moon in this decade and do the other things, not because they are easy, but because they are hard.' In retrospect it wasn't that difficult: all that was needed was Newton's laws of motion and a rocket. But still, the rocket had to be made and tested, and that was part of the Gemini mission.

Gemini 8 launched on 16 March 1966, with a planned

rendezvous with an unmanned spacecraft called Agena and the first-ever spacewalk. Armstrong managed to dock with the Agena, but then the two craft began to roll. Armstrong undocked but the roll increased to the point where the craft was rotating once per second. This could have been fatal, as the astronauts on board could have passed out. He managed to bring the craft under control and was rewarded with the NASA Exceptional Service medal, and also a $678 pay rise. This mightn't seem like much, but it raised his pay to $21,654 (around $172,000 today), making him NASA's highest-paid astronaut (which presumably irritated the hell out of the other astronauts).

In late 1968 came the news that Armstrong, along with Buzz Aldrin and Michael Collins, would be the crew of the Apollo 11 mission to land a man on the moon. Armstrong was almost killed in a test flight of a lunar landing vehicle but ejected on time, and later said the experience really helped him when it came to the actual landing. Records show that Armstrong was chosen to be the first one to walk on the moon because NASA management felt he didn't have a large ego and so the experience mightn't go to his head (unlike his blood, which most certainly did), but at a press conference on 14 April 1969 they said the reason was the design of the landing module, which put Armstrong nearest the opening hatch.

On 16 July 1969, the Saturn V rocket launched Apollo 11 from the Kennedy Space Centre. Von Braun, who had helped develop the Saturn V, said of the event: 'I certainly prayed a lot before and during the crucial Apollo flights.' He had undergone a religious conversion to evangelical Christianity, and explained, 'The farther we probe into space, the greater my faith.' One has to wonder what else went through his mind. Did he think about his journey from designing rockets for the Nazis to this point – a rocket that would bring men to the moon?

During the launch, Armstrong's heart rate peaked at 110 beats per minute, more than double his resting heart rate. The rocket safely reached the moon, and the Eagle landing module was launched. During the landing, several computer-error alarms sounded but the crew were told by mission control in Houston not to be concerned. Armstrong noticed they were heading to a landing zone that looked unsafe and so he took over manual control. When the Eagle, er, landed, there was only 20 seconds of fuel left, so they just about made it. Armstrong, cool as can be, announced to mission control and the world possibly the second-most famous words in science: 'Houston, Tranquility Base here. The Eagle has landed.'

Armstrong and Aldrin celebrated with a firm handshake and a pat on the back. What good chaps! Mission control replied: 'Roger, Tranquility. We copy you on the ground. You got a bunch of guys about to turn blue. We're breathing again. Thanks a lot.' During the landing Armstrong's heart rate got as high as 150 beats per minute.

The crew were then scheduled to rest, but Armstrong wanted to crack on and moved the moonwalk forward. At the bottom of the ladder, he uttered the most famous words in science: 'That's one small step for man, one giant leap for mankind.' He had thought of this phrase just before he left the lunar module. Later he said this was because he had put the chances of the mission succeeding at 50/50: 'It didn't seem to me there was much point in thinking of something to say if we'd have to abort the landing.' This was disputed by his brother Dean Armstrong (who must have been dead jealous), who said Neil had shown him a draft of the lines months before.

One issue that is still disputed is whether Armstrong said 'one small step for *a* man', because if he didn't the line doesn't make sense, as 'man' and 'mankind' are equivalent. For years Armstrong said static had obscured the 'a', but he later conceded that he must have dropped it, asking that history grant him 'leeway'.

Whatever the truth of the matter, an estimated 650 million people heard him and watched the Apollo 11 moon landing. That's 14.6 per cent of the world's population. I remember myself being five years of age, and my father calling me in from playing to say, 'Watch this!' On the screen of the black-and-white TV in the corner of our living room, I saw Neil Armstrong and Buzz Aldrin on the moon. I spent the next five years drawing rockets and spaceships on the covers of my schoolbooks. (The third man to walk on the moon, Pete Conrad of the Apollo 12 mission, didn't mess up when it was his turn. He said, 'Whoopie! Man, that may have been a small one for Neil, but that's a long one for me.' He also had a great motto: 'If you can't be good, be colourful.')

Aldrin joined Armstrong 19 minutes later. Quite what he was doing for those 19 minutes isn't known. Polishing his space helmet for the cameras? Armstrong unveiled a plaque while Aldrin planted the flag of the United States. The US had, after all, paid for all this. The programme had cost an estimated $28 billion, which is the equivalent of $230 billion today – a mind-boggling sum of money. The flag had a metal rod in it to make it look like it was flying. Armstrong's preference was for it to be draped on the flagpole without the metal rod, but he was overruled.

After speaking to President Richard Nixon, Armstrong

went on a brief walk. Aldrin left a package of memorial items to the Soviet cosmonauts Gagarin and Komarov, and the Apollo 1 astronauts who had died in a fire during training. They had also brought with them two fragments of wood from the propeller and four pieces of fabric from the wing of the 1903 *Wright Flyer*, the first airplane, though they didn't leave them on the moon. They didn't bring any of the paper from the Montgolfier brothers' balloon.

In 2018, Armstrong's sons put a lot of their father's memorabilia up for auction, and the sales totalled $7.5 million. The fragments from the *Wright Flyer* sold for between $112,500 and $275,000. All three astronauts – including Armstrong – were actually poorly paid for the mission to the moon, even though Armstrong was on a good yearly salary. When they went to the moon, they received $8 per day ($65 in today's money). This was the same as they received back at base. And various deductions were taken from that pay, including accommodation in the Apollo 11 spacecraft.

They then climbed back on board and prepared to lift off. However, Aldrin somehow managed to hit and break the ignition switch for the engine with his backpack. In the words of Homer Simpson: 'Doh!' Using a pen, they pushed the circuit breaker to start the launch sequence. If they hadn't used that small, felt-tipped pen, the mission could have ended in disaster. It's on display in the Museum of Flight in Seattle if you want to see it. The Eagle then rendezvoused with the command module Columbia, reuniting the three astronauts. Michael Collins was referred to as the loneliest man in existence for his solo mission spent orbiting the moon, at times out of contact with his fellow astronauts and the Earth. He took a picture at one point which had the moon and the Earth in it. All of humanity bar himself was in that picture.

On their return to Earth, rather ridiculously they had to go through customs and immigration. Red tape. They then went into an 21-day quarantine (just in case they'd picked up an alien infection, even though they hadn't seen John Hurt in the movie *Alien*) and then went on the 38-day 'Giant Leap' world tour, which began with two litter-filled ticker-tape parades in New York and Chicago, attended by 6 million people. There was a state dinner in Los Angeles attended by representatives of 83 nations. The Apollo 11 mission was billed as a success for all humanity, which indeed it was. Armstrong then went to Vietnam with comedian Bob Hope to entertain the troops. After Apollo 11 he said he would never fly in space again, in order to give others the chance. His last act of exploration was in 1985, when he went with a group of the 'greatest explorers' to the North Pole. The first person to summit Everest, Edmund Hillary, was in the party, and Armstrong said he only went because he'd seen the North Pole from space and wanted to see it from the ground.

Armstrong's family called him a 'reluctant American hero', and he led a relatively quiet life after all the fuss died down, becoming a college professor of aerospace engineering for a while, and getting involved in several businesses. Looking back on his life, he said: 'The one thing I regret was that my work required an enormous amount of my time, and a lot of travel.' They had at least been reimbursed for travel. The voucher documented their journey: 'From Houston, Texas to Cape Kennedy, Fla., Moon, Pacific Ocean (USN Hornett), Hawaii and return to Houston, Texas. Net to traveler: $33.31.'

One of the more notable things that happened to Armstrong was in 2005, when he had a dispute with his barber of 20 years. After cutting his hair, the barber sold some to a collector for $3,000. Armstrong insisted that he either retrieve the hair or give the money to charity. The barber was unable to obtain the hair and so gave the money to charity.

Valentina Tereshkova

Women are conspicuously absent from the standard NASA story. Fortunately, this has been corrected somewhat with the recognition of a group of women mathematicians who worked at NASA all through the space programmes. Like the women we saw in the last chapter who were tasked with mapping the stars, they were called 'computers'. Their job was to calculate flight trajectories, with Katherine Johnson, Dorothy Vaughan and Mary Jackson all playing significant roles. Johnson was important to the Alan Shepard, John Glenn and Apollo 11 missions, and also worked on the Space Shuttle programme. She was also notable as African Americans were few and far between at NASA at the time. She captured the excitement of the space programme when she said: 'Everything was so new – the whole idea of going into space was new and daring. There were no textbooks, so we had to write them.' She also said: 'Girls are capable of doing everything men are capable of doing. Sometimes they have more imagination than men.' In 2015 she was presented with the Presidential Medal of Freedom.

The Soviet Union was more supportive of women in their space programme and in the workplace more generally. Valentina Tereshkova flew a solo mission into space on 16 June 1963, orbiting the Earth 48 times. At that point, Russia were four-nil up in the space race: Sputnik, animal in space, man in space and woman in space. Tereshkova had been a factory worker but developed an interest in skydiving as a hobby. Nikolai Kamanin, the director of the cosmonaut training programme, had read that in the US women were training to be astronauts. There was a programme called Mercury 13, with an all-women roster. They proved as capable as the men, passing all the tests, but Congress and specifically President Johnson shut the programme down, saying, 'Let's stop this now!' But Kamanin didn't know that, and said: 'We cannot allow that the first woman in space will be American. This would be an insult to the patriotic feelings of Soviet women.' It took until 1983 for an American woman to go into space, when Sally Ride flew in the Space Shuttle.

Tereshkova applied to join the programme and learned how to fly a MiG-15UTI jet fighter as part of her training. Once she was selected, sadly Kamanin referred to her as 'Gagarin in a skirt'. After her mission into space, she had logged more flight time than the combined times of all the American astronauts at that point. To celebrate her success, 1 million flowers were put on display and she was awarded a 'Hero of the Soviet Union' medal. She married another cosmonaut, Andriyan Nikolayev, on 3 November 1963, with Soviet premier Khrushchev in attendance. Kamanin described the marriage as 'probably useful for politics and science'.

Space Oddity

Once the entire Apollo programme had ended, interest in space travel diminished. The media hardly covered the Apollo 18 mission at all. The expense of the missions was an issue, since it wasn't fully clear what they actually achieved, other than showing the Russians the might of US technology. However, they would have a significant cultural impact, and are estimated to have inspired 1,800 spin-off products – from integrated circuits in computers to fireproof material to heart monitors, and even Teflon coating for saucepans. The Apollo programme also stimulated environmental activism, driven by the images of the Earth taken from space, which emphasized the fragility of our only home.

Future Apollo missions were cancelled, but NASA kept lobbying Congress for funds, concerned for the 400,000 workers involved in the Apollo missions. The redoubtable von Braun got involved and began pressing for an orbiting space station, and the idea of Skylab was born. It was launched on 14 May 1973, and its crew were deployed there in three separate missions. Several experiments were carried out in Skylab, from measuring human physiology in space, to work on solar physics and astronomy and the imaging of the Earth to monitor land and vegetation patterns. It was left unoccupied and fell to Earth in July 1979, and the multimillion-dollar spacecraft was destroyed on impact. This makes writing off an old car seem trivial.

Construction of the Space Shuttle got underway in 1974, and it flew from 12 April 1981 until 11 July 2011, in 135 missions. Sadly, two of these missions didn't return – the Challenger and Columbia disasters, which were caused by equipment malfunction. Each flight was estimated to cost $260 million. Altogether, the programme cost $196 billion.

The biggest programme that NASA is currently involved in is the International Space Station. This is a multinational collaboration involving five space agencies, including the European Space Agency. Astronauts are carrying out experiments in astrobiology, astronomy, meteorology and other fields. One of its most memorable events was when commander Chris Hadfield shot a music video of David Bowie's 'Space Oddity' while aboard, the first music video ever filmed in space. NASA also continues to send unmanned missions to explore our solar system. These are often named after the famous astronomers described in the last chapter: for example Galileo (going to Jupiter) and Cassini-Huygens (to Saturn).

Better Space Toilets

In recent years the focus has shifted increasingly to private space travel, which is supported by two of the world's richest men, Jeff Bezos and Elon Musk. It's an interesting turn of events that two men have made so much money on Earth that they are now funding programmes to escape it. And it can't just be for tax purposes, as the IRS doesn't consider space as being abroad for US citizens.

In 2002 Musk founded SpaceX, a space transport service, having made most of his money with the online payment service PayPal. The story of SpaceX begins with Musk travelling to Moscow in 2001 to buy some intercontinental ballistic missiles with the intention of using them to fly in space. He was eventually turned down amid reports that one of the chief designers of the Russian rockets spat at him. Musk then began to build his own rockets, and after three failed launches he finally had a successful launch with Falcon 1 in 2008. Later that year, SpaceX signed a contract worth $1.6 billion for 12 delivery flights to the International Space Station. With the Space Shuttle out of commission, another transportation system was needed, and SpaceX filled that gap. SpaceX is now developing Starlink, a constellation of satellites around the Earth, to provide worldwide internet access.

Meanwhile Jeff Bezos, who made his money from Amazon, has a space-flight company called Blue Origin. He had been interested in space travel and space tourism for some time. In 2015 he announced New Shepard, an orbital launch vehicle. On 20 July 2021, New Shepard took four people into space: Jeff, his brother Mark, a friend Wally Funk (who was part of NASA's Mercury mission all those years ago) and a Dutch student named Oliver Daemen, whose father paid $28 million for the privilege. Funk and Daemen became the oldest and youngest people to travel into space, and the mission was the first space flight with civilians. Just like Yuri Gagarin's flight, the first-ever into space, it crossed the Kármán line – the boundary between Earth's atmosphere and space – and was in space for 10 minutes.

New Shepard is named after Alan Shepard, and Bezos is building another spacecraft which he will call New Glenn, after John Glenn. He is obviously proud of the legacy of NASA, and funded the recovery of two Saturn V engines from the floor of the Atlantic Ocean. These were from the Apollo 11 mission that brought Armstrong, Aldrin and Collins to the moon in July 1969. However, his love of NASA didn't prevent him from embarking on litigation against the agency. He sued them for declining to offer a contract to Blue Origin for the next mission to the moon involving humans. He claimed that NASA didn't evaluate his lunar-lander proposal, and instead offered the contract to rival company SpaceX – which must have stuck in Bezos's throat. NASA want to return humans to the moon as soon as 2024 – their last such mission being Apollo 18 in 1972. Blue Origin offered a contract worth $5.9 billion, but NASA went with SpaceX – who were charging $2.9 billion, a mere snip, and were presumably the lowest bidder (as with previous missions).

Whatever happens, NASA is committed to putting humans on the moon and is pledging that the crew will have a woman and a person of colour on board. As of 2022, 75 women have flown in space (compared to 600 men), but women will play an increasing role in space travel, particularly as there is evidence that they might be better suited to longer space missions. Studies have shown that they are more mutually supportive and create a more positive environment. Spacesuits were actually made from a material used to make bras, but until recently insufficient numbers of smaller spacesuits were made. Things have changed somewhat though. In 2007, Peggy Whitson became the first woman to command the International Space Station. In 2019, the first all-female spacewalk occurred, involving Jessica Meir and Christina Koch. And then, real progress. In October 2020, a toilet with a better design for women was installed on the International Space Station. The seat is tilted and the toilet is taller. Melissa McKinley, a project manager at the Johnson Space Centre, has said it should help the astronauts better position themselves for a 'No. 2'.

The most recent NASA mission to the moon is Artemis, named for the Greek goddess of the moon and twin sister of Apollo. NASA plans to land humans on the lunar south pole in 2025 and create a sustainable base camp by 2028. There may be ice near the south pole in a crater called (appropriately enough) Shackleton after another great explorer – one who went to Antarctica in the early 1900s. The first flight of Artemis in November 2022 was captained by a crash-test dummy called Commander Moonikin Campos (named after the engineer who saved Apollo 13 from disaster), and it flew to the moon before orbiting and returning to Earth. This first flight was also used to test new spacesuits as well as a voice-activated digital assistant, hopefully more like Alexa than HAL the malevolent computer in *2001: A Space Odyssey*.

NASA has said the mission is about establishing a long-term presence on the moon, to prepare for what it calls the next 'giant leap': sending the first astronauts to Mars. NASA has also stated 'we are closer to landing crew on the moon again than at any other time in our history since the Apollo program. The sooner we go to the moon, the sooner we send astronauts to Mars.' And you never know – when that person walks on Mars, they may say, 'One small step for a woman, one giant leap for humankind.'

3. Get Your Rocks Off

Travelling into space was fun. It might even have been the best fun that science ever made happen, hurtling towards the moon at a top speed of 1 kilometre per second, with the Earth receding in the rear-view mirror. It really appealed to our spirit of adventure. And it gave us a whole new perspective on our planet that the pure facts do not convey. Tell me that the Earth is a 12,742-kilometre-diameter sphere of solid rock with a dense metal core, with areas of surface water around 3.7 kilometres deep and solid regions overlaid with organic material (microbes, plants and animals), covered with a gaseous atmosphere that extends a further 12 kilometres above the surface and – well, I'm impressed by the numbers but *Zzzzzz*. However, show me a colour photograph of a stunning blue jewel hanging in the pure blackness of space and *WOW*, my emotions are firing like a rocket. Now, to make a real impact, put the science and beauty together. The iconic photograph of Earthrise from the moon, taken on Christmas Eve 1968 by Apollo 8 astronaut Bill Anders, clearly shows the difference that thin layer of gas and water (less than 0.1 per cent of the Earth's diameter, where all life resides) makes, compared to the stark, lifeless rock of the moon. Science had already been generating the facts about the relative fragility of our climate and environment, but being able to view our planet from the outside thanks to the Apollo missions is seen by many as the catalyst for the burgeoning area of environmental sciences.

I almost became a geologist myself. When I went to university to study science, I hadn't quite made up my mind as to whether I wanted to be a geologist or a biologist. When I was a teenager, I found the fact that there were so many different types of rocks very interesting and tried to collect as many as I could. I grew up near the Wicklow Mountains in Ireland, which was a source of different rock types – mainly granite but also rocks like quartzite and schist (a name that still makes me chuckle, because like my enthusiasm for science, my sense of humour hasn't changed much since I was a boy). I had boxes full of different rocks under my bed. My mother thought I was mad. A lot of scientists start out as collectors. We like to classify things, probably because it gives us comfort. I've heard from psychologists that one way to deal with overthinking (and don't we all do that from time to time, especially scientists?) is to write down your thoughts, or make a to-do list, or classify things. It provides structure for our thoughts and the world around us. It takes the enormity out of existence, which can be overwhelming. Unless you're like Walter in *The Big Lebowski*, whose default was always to go bowling.

On my first day at university I went to one lecture on geology and one on biology. Both lectures contained facts but geology was *Zzzzz*, and biology was more *WOW*. It is on these whims that careers are made.

The First Rock Star

Earth science inevitably began with the Greeks. All that sitting around in the sun and being well fed must have given them time to ponder rock. Theophrastus was a philosopher who wrote a book with a very simple title: *On Stones*. Despite what Mick Jagger might think, there's no doubt what that book was about, right? Theophrastus had the unenviable task of taking over the school that had been founded by Aristotle 13 years previously, so he had to be really smart. Aristotle had left him his library and designated him his successor, giving rise to the usual bickering – two of his colleagues, Eudemus and Aristoxenus, complained. Theophrastus probably told them to be philosophical about it.

He wrote books about many things, including zoology, physics, poetry and ethics. He wrote a particularly important book on botany called *Enquiry into Plants* (somewhat disappointingly, not *On Plants*). In *On Stones*, he described minerals and ore from the mines near Athens. He also described precious stones like emerald, amethyst and onyx, and knew that pearls came from oysters. *On Stones* was a standard text on geology used right up to the Renaissance, and Theophrastus was the original rock star of geology, being the first to systematically classify different kinds of rock. He did this based on their behaviour when they were heated, which was a scientifically clever thing to do as he could measure different properties and classify accordingly. Remember, science is about collecting data and interpreting it. Theophrastus observed that 'some can be burnt, whilst others can be melted and then there are those which just break up into smaller pieces'. He classified obsidian and amethyst together because they both change colour when heated. He also grouped minerals together based on common properties, for example amber and magnetite, which both attract lightweight materials like feathers when rubbed. Theophrastus died at 84, having said 'we die just when we are beginning to live' – in contrast to Roger Daltrey who at 21 famously sang that he hoped he'd die before he got old. At the time of writing, Daltrey is 79.

A friend of mine, Brian McManus, in something of a mid-life crisis, is following in the footsteps of Theophrastus. He has taken to collecting rocks and gemstones and has even bought a 'tumbler', which is a device for polishing rocks. His is a house that is often full of the sound of tumbling rocks. His wife is beside herself. Theophrastus would be amazed at the ease of rock identification nowadays. Using an app, Brian can scan and quickly identify online anything he finds. He has dragged me out on gemstone-hunting missions on beaches in Ireland, but every rock we find disappointingly comes up as jasper. I had to look up what jasper is. It's defined as 'an aggregate of microgranular quartz and/or cryptocrystalline chalcedony'. Being a scientist, I tested the technology, and scanned in a digestive biscuit, which came up as . . . jasper. As mentioned previously, this is what scientists call a negative control, designed to make sure the method of measuring something is reliable. If everything comes up jasper, then the method of identification isn't working properly. Unless *The Hitchhiker's Guide to the Galaxy* has it wrong and the number 42 isn't the meaning of life – actually, it's jasper.

Comte de Buffon

In the eleventh century, Shen Kuo, a Chinese naturalist, wrote what is now seen as a prophetic book. It was one that was important for geology, because unlike the work of Theophrastus – which simply classified existing rocks – it speculated about how rocks formed in the first place. Shen wondered why there were marine fossils in the Taihang

Mountains hundreds of miles from the Pacific Ocean, and concluded that mountains were formed by uplift, and the land was formed from erosion by the elements and the laying-down of silt. As we will see in a later chapter, Darwin wondered about something similar when he observed the fossils of shells high up in the Andes – except Shen Kuo was writing eight centuries earlier. Like Theophrastus, Shen contributed to many fields. He even gets the credit for the first description of the magnetic needle compass, which would be hugely useful for navigation. This wouldn't be known about in Europe for another 400 years.

Shen wrote about how the formation of mountains must take place over vast periods of time. He hypothesized that climate change was happening, by reporting on fossilized bamboo. He knew that where the bamboo was found could no longer support the growth of bamboo, and so the climate must have changed.

In Europe, early geologists tried to find evidence for the Great Flood, since the Bible was considered a true and accurate account of history. In medieval times, the Bible was constantly read, reread and analysed. There was no Netflix then, nor the opportunity of watching constant reruns of *Game of Thrones*. Most people who read the Bible just accepted it as holy writ, but it also got other people thinking and puzzling over things, not least the origin of the Earth.

Because it was believed that the Great Flood had happened, evidence was sought for it. This mainly concerned the layers (called strata) that were observed in rocks. It was thought that the Great Flood had somehow laid down those strata. This conclusion quickly became untenable, however. More and more descriptions were published of the fossils of all kinds of exotic-looking creatures that were no longer alive on Earth. It wasn't at all clear where they came from unless the Great Flood had somehow wiped them all out. The species that didn't make it when Noah gave out the call. Not unreasonable at the time I suppose, and certainly a potentially comforting answer. But science should never be about being comfortable. It should be about the truth.

In the eighteenth century, French naturalist Georges-Louis Leclerc (whose aristocratic title was Comte de Buffon) attacked the biblical account of how the Earth was formed and the Great Flood. He speculated in his 1749 book *Histoire Naturelle, Générale et Particulière* (yet another great book title, which I would have liked to use for this book) that the Earth was 75,000 years old, based on how quickly globes cool. Just like when Copernicus wrote about the Earth going around the sun, this was derided at the time. James Ussher had previously calculated from the Bible that the Earth began on 22 October 4004 BCE at around 6 p.m. Presumably God had to get this done ahead of the children's bedtime.

Ussher was the Primate of All Ireland ('primate' in this context doesn't mean monkey). Born in Dublin in 1581, his early education was thanks to his two blind spinster aunts. His calculation for when the Earth was created was accepted for the next 200 years and even appeared in the King James Bible. Famous evolutionary biologist Stephen Jay Gould gave Ussher credit for his efforts, as he at least took a systematic approach to analysing the genealogy of Jesus Christ as it was written in the Bible. Ussher displayed great scholarship to justify the date he arrived at. He cited a lot of ancient history – of the Persians, Greeks and Romans. He also described ancient calendars and chronologies. He mentioned the death of Alexander the Great in 323 BCE and of Julius Caesar in 44 BCE and discussed the Babylonian king

Nebuchadnezzar. He cross-referenced the Torah (the first five books of the Hebrew Bible), calculating that Solomon's Temple was built in 1012 BCE, a thousand years before the birth of Christ – who was thought to be the 'fulfilment' of the temple.

Ussher's work was seen as a tour de force of scholarship and learning, and while its accuracy is debatable, at the time it was considered highly believable. The majority of people in Europe at that time therefore accepted that the Earth had been formed at 6 p.m. on 22 October 4004 BCE. And that was that. Yet again, comforting in its own way I suppose. But almost from that moment on, geology would put a huge spanner in that particular Earth works.

Sucking Chocolate Nuts

The mining industry began to grow in the sixteenth century, and this allowed for much debate on types of rocks and when they might have formed and how. Charles Lyell wrote a very influential book called *Principles of Geology* which presented evidence that geological change was very slow and gradual and ongoing. Much slower than was possible if the Earth had formed less than 6,000 years ago. Up to that point, it was generally accepted that the Earth had been formed in an instant and hadn't changed much since then.

Lyell was originally a lawyer, and spent some time in Lincoln's Inn, London, in the early 1820s – curiously following in the footsteps of Ussher, who had been a preacher there. And I followed in both their footsteps by studying for my PhD in Lincoln's Inn Fields, although I hadn't yet heard of Lyell or Ussher when I was there in the 1980s. Lyell then became a travelling lawyer so that he could observe geological phenomena wherever he went, and in 1827 he left the legal profession to become a full-time geologist. He spent his honeymoon on a geological tour of Switzerland – not as unromantic as it sounds, as his wife, Mary Horner, shared his interest in geology and contributed to her husband's scientific work. It's not known whether they got their rocks off in Switzerland.

Lyell visited the United States and Canada in the 1840s, and after the Great Chicago Fire in 1871 when libraries were destroyed, he donated many of his books to help found the Chicago Public Library. Mount Lyell, the highest peak in Yosemite National Park, is named after him. He gets credit for being 'the spiritual saviour of geology, freeing the science from the old dispensation of Moses'. Ussher must have been turning in his grave.

The big idea in *Principles of Geology* is uniformitarianism, a big word meaning that the Earth is entirely formed from slow-moving forces still in operation acting over enormous periods of time. This is in opposition to catastrophism, where big changes happen abruptly and unexpectedly. One justification for catastrophism at the time was that valleys often have small rivers in them, and it was thought this must have happened because of a rapid change in the size of the valley. It began small, with a small river running through it, but then something catastrophic happened – say a massive earthquake – which somehow made the valley much bigger without changing the size of the river. Lyell didn't like catastrophism, saying: 'Never was there a dogma more calculated to foster indolence, and to blunt the keen edge of curiosity.'

In 1858, Lyell also organized the co-publication of the

theory of natural selection by Darwin and Alfred Russel Wallace (as we will see later) – despite his personal difficulties in reconciling evolution with his religious beliefs. Darwin had gone on a voyage of discovery with Robert FitzRoy on board the most important ship in science (after the starship *Enterprise*, of course), HMS *Beagle*. FitzRoy gave Lyell's book to Darwin, who became a fan and speculated on the process of uplift.

Although Lyell didn't try to date the Earth, he knew that it was very, very old. In 1864 William Thomson, 1st Baron Kelvin, speculated that the Earth was somewhere between 20 million and 400 million years old – based on the cooling of a molten object. He was famous for work on temperature, and the temperature scale called Kelvin is named in his honour. He managed to determine a temperature called absolute zero – which is the lowest temperature that can possibly be reached, hence 'absolute'. And he wasn't talking about vodka.

Kelvin was also interested in the laying of a transatlantic telegraph cable, and on one cable-laying expedition he spent time in Madeira, becoming friends with one Charles Blandy and his three daughters. On a subsequent visit he asked the Blandy residence, 'Will you marry me?' – not indicating which of the three daughters was his intended wife. It looks like Frances Anna Blandy responded first, and they were married.

Kelvin was intrigued by the age of the Earth, and although he was a creationist, he was among the first to conclude that the Great Flood couldn't explain everything, and he didn't count himself among what were at that time called flood geologists. He finally settled on between 20 million and 400 million years old, which at the time was seen as ridiculously old.

By the twentieth century, radioactive decay was used to

calculate the age of rocks. A friend of mine recently told me that at Christmas if she gets a chocolate with a nut in it, she will suck the chocolate from the nut, but then return the nut to the wrapper, putting it back into the box of chocolates to surprise her family. Let's say she does this every hour. If you count the number of nuts-in-wrappers relative to the ones still in chocolate, you can figure out how long it has been since she took the first chocolate. This is how dating things by radioactive decay works.

There are over 300 naturally occurring isotopes. An isotope is defined as two or more forms of the same element that contain equal numbers of protons but different numbers of neutrons – protons and neutrons being subatomic particles. Some isotopes don't change over time, but others are unstable and break down into other isotopes by a process called radioactive decay. The decay happens in the nucleus of the atom, which releases various particles – for example, neutrons. This eventually stabilizes the nucleus.

Carbon-14 is an isotope of the element carbon with six protons and eight neutrons. The more common isotope is Carbon-12, with six protons and six neutrons. It's stable and doesn't decay. But Carbon-14 does. Over time it will decay into Nitrogen-14, by emitting subatomic particles. This decay can be used for the carbon dating of biological materials – or, using other elements, for dating rocks. The 'parent' atom decays into a 'daughter' atom at a stable rate. Physicists talk of 'half-life', which is the time it takes for half a set amount of the parent atoms to decay. Carbon-14 has a half-life of 5,730 years, which is the length of time it takes for half an amount of Carbon-14 to decay into Nitrogen-14.

Carbon-14 is also very useful in biology experiments, because it can be used to mark proteins. If you feed a mouse food containing Carbon-14, it will mark all the proteins, since the Carbon-14 is used to make proteins in the mouse's body. Proteins are made of amino acids which are carbon-rich (see chapter 5). We take proteins in as part of our diet, but then break them down into amino acids to use to make more proteins. If the amino acids in the diet are labelled with Carbon-14, that will end up in new proteins, which can then be measured in the body.

When I was a PhD student in the aforementioned Lincoln's Inn Fields, I was working with Carbon-14 to track a protein in the immune system called cyclo-oxygenase, which is involved in the inflammatory process. Cyclo-oxygenase promotes inflammatory disease, particularly in the case of arthritis, and I was able to measure the level of the protein going up during inflammation. The department I was in had an annual allowance for how much radioactivity we could work with, since radioactive elements need to be handled carefully as they can cause cancer. One year, I ordered the whole year's allowance for radioisotopes, because I thought my experiment was so important. Sadly, the experiment didn't work, but I always made sure to protect myself from radiation. A special lead apron had to be worn, as lead is protective against radioactivity. A French scientist I once worked with was especially paranoid, and also wore a special protective garment involving two pouches which were placed over the testicles. He called them his *boules testicules*. Strange how some memories stay with you.

Using the decay of uranium and lead isotopes, Arthur Holmes calculated that some rocks from Norway were 370 million years old. This calculation wouldn't have been possible without Pierre Curie, Marie Skłodowska Curie and Henri Becquerel's work on radioactive elements, which we'll read about in chapter 4. It's a great example of cross-fertilization between different disciplines, as the Curies' discovery in chemistry was used in geology.

We were getting closer to the actual age of the Earth, and it was Holmes who championed the use of radioactivity to age the planet. Holmes then identified rocks from Ceylon as being 1.6 billion years old. Things were getting much older. By 1927, he had revised the figure further to 3 billion years old, and finally came up with an age of 4.5 billion years – which is still the agreed age of the Earth.

The fact that our planet is billions of years old is still amazing to me. It's certainly a stark contrast to the 6,000 years old that the Bible seemed to suggest, and which – as recently as 2014 – 40 per cent of Americans still believe for reasons that escape me, given the irrefutable evidence, based not least on the radioactive dating of rocks. Science tells us we live on a planet that is 4.5 billion years old, in a universe that we saw in chapter 1 is 13.799 billion years old and gazillions and gazillions of kilometres (actually 93 billion light years or 8.8×10^{23} kilometres) across. Unfathomable space and time. And yet we still worry about what we might have for dinner.

Finding the Curve

Anti-science can prevail in geology too, and there are even some people out there who still think the Earth is flat (although I have never yet met one). This is perhaps the best example of the difficulties scientists can have. They can show all the evidence there can possibly be of something (like the Earth not being flat), but people still refuse to believe it. This tells us something about how humans are. If they decide something is the way it is, and if they are part of a group or movement who believe the same thing, it can be hard for them to back down, even in the face of evidence that can't be denied. They can end up cutting off their noses to spite their faces.

The International Flat Earth Research Society was set up by Samuel Shenton in 1956 and was initially seen as an example of English eccentricity. In the 2000s, because of the internet, the idea began to spread again. Most flat Earthers think the sun moves in circles around the North Pole. The most recent 'US model' claims that the sun and moon are 50 kilometres in diameter and circle the Earth at a height of 5,500 kilometres, whereas the real numbers are a diameter of 12,742 kilometres for the Earth and 3,474 kilometres for the moon, which orbits the Earth at a distance of 357,000–407,000 kilometres. One survey found that 7 per cent of Brazilians believe the Earth is flat. This has been put down to an expanding evangelical Christian movement there.

In 2017 the US rapper B.o.B (real name Bobby Ray Simmons Jr) started a campaign to fund a satellite to find evidence that the Earth was a disc. His aim was to 'find the curve', a term flat Earthers use to describe the edge of what they see as our disc-shaped planet. There has been a worrying rise in the number of people who believe in flat Earth theories. There's even an annual flat Earth conference in the US, and the most recent (in 2018) was attended by 650 people. Of course, in a country of 332 million, that isn't many, but to put it into context, around 4,000 people attended the last immunology conference I went to in the US. It's not that big of a difference. The people attending the flat Earth conference were reported as being sincere in their beliefs. There were lectures entitled 'Talking to your family and friends about flat Earth', 'NASA and other space lies' and '14+ ways the Bible says the Earth is flat'. Despite the challenges, we can learn a lot from this. It could even help in our efforts to understand why people deny climate change, or think that vaccines are no safer than the diseases they prevent, or believe that drugs like Ivermectin can treat COVID-19 (as we'll read about in chapter 7).

Physicists have attempted to counter flat Earth views with evidence, and are trying to come up with ways to fight back. Lee McIntyre went to the recent flat Earth conference to find out more about why they think the way they do, and he included his experiences in his book *How to Talk to a Science Denier: Conversations with Flat Earthers, Climate Deniers and Others Who Defy Reason*. He noticed that all the flat Earthers he spoke to also didn't believe that humans had walked on the moon. That particular conspiracy theory emerged soon after the moon landings. I am a fan of the movie *The Shining*, which was directed by Stanley Kubrick (who also made *2001: A Space Odyssey*), and some moon-landing conspiracy theorists claim that Kubrick was hired to film the fake moon landing. They say that he left clues admitting to his involvement in *The Shining*.

The film involves a family spending a winter as caretakers in a remote hotel, where the father – played by Jack Nicholson – goes insane and tries to kill his wife and son. The supposed 'evidence' is the pattern on the carpet in the hotel resembling the launch pad of Apollo 11. The son, Danny, rises off the carpet with a toy rocket, wearing a jumper which depicts the Apollo mission. The twins that appear apparently represent the Gemini mission – the book the film is based on only had one child. It is said that Kubrick changed the haunted room number to 237 because the distance to the moon is 237,000 miles. Kubrick fans say he was just having fun.

McIntyre also recorded that all the people he spoke to had

been hugely influenced by social media. The internet has helped to normalize the flat Earth theory – and indeed other conspiracy theories – and it also feeds into confirmation bias, which is a tendency to interpret new evidence as confirming what you already believe.

A key piece of evidence that the Earth isn't flat and that it rotates is the Foucault pendulum, named after the French physicist Léon Foucault. On 3 February 1851, Foucault, who had dropped out of medical school, assembled a group of scientists, inviting them 'to see the Earth turn'. He hung a heavy weight from a 67-metre-long chain in the Meridian Room of the Paris Observatory. The pendulum swept through the air, but as time went by it changed direction. This was obvious as the tip of the pendulum was scoring lines in the sand that had been placed under it. As it was hanging from a fixed point, the only way to explain this was that the Earth was turning under the pendulum. You could create this experiment yourself by taking a beach ball, putting wet sand over the top of it, then swinging a pendulum through the sand as you turn the beach ball. The turning of the beach ball will lead to the direction of the lines in the sand changing – provided the wet sand doesn't fall off it, of course. Maybe jam would be better. He debated the experiment in the Pantheon, a beautiful former church and mausoleum in Paris where both Pierre and Marie Curie's remains are buried. The pendulum changes direction over the course of the day as it swings, providing direct evidence of the Earth's rotation.

Flat Earthers claim that Foucault's pendulum experiment was fraudulent and manipulated by magnets. McIntyre's conclusion is that the best way to win over people with these kinds of beliefs is to 'restore trust in scientific organisations and institutions' – but that's easier said than done.

Men of Learning Have Sucked Her Brains

Some creationists believe that fossils were placed in rocks to challenge faith. Very unlikely, but try convincing them otherwise. Fossils have always been interesting, especially one set of fossils above all: the dinosaurs. The Chinese thought they were from dragons, which is where that legend might have come from. In the early 1800s, geologists in England began describing dinosaur bones, considering them to be 'great fossil lizards'. The term 'dinosaur' was coined by fossil specialist Richard Owen.

Fossil hunting became very fashionable in the nineteenth century. Mary Anning was one such collector, and she discovered the fossils of many different dinosaurs, including the ichthyosaur and pterosaur. She was born in Lyme Regis in Dorset in 1799; her parents had ten children, but only two survived into adulthood. Most of the others died of infectious diseases, which wasn't uncommon at that time because of overcrowding and a lack of effective treatments or vaccines. Smallpox and measles were very common causes of death. Try telling that to anti-vaxxers.

As a 15-month-old, Anning was struck by lightning while being held by a neighbour under a tree. The neighbour and two other women beside her were killed. Anning survived, and people would attribute her curiosity and intelligence to this incident. (Breaking news: the science is now in, and it's clear that being struck by lightning is less effective than hard work and a good education as the basis for a successful career in palaeontology.)

Lyme Regis had become a popular tourist destination and many locals made money from selling 'curios'. These were colourful fossils with names like 'snake stone' and 'devils finger', because they looked like a coiled snake or a finger. These fossils had been laid down in what was then a shallow sea bed, in the Jurassic period (about 201–145 million years ago). Anning's father supplemented his income as a carpenter by mining the local cliffs for fossils, and she would join him. Her first major find was a four-foot ichthyosaur skull. Ichthyosaurs are fossil marine reptiles that resemble dolphins. It was called a 'Crocodile in a Fossil State' and generated tremendous interest. It was bought from Anning for £23, a huge sum of money equivalent to £2,500 today. At the age of 27 she opened a shop, which attracted geologists and fossil collectors from all over the world. She supplied fossils for the newly opened New York Lyceum of Natural History.

Because she was a woman, she didn't receive credit for her important discoveries. An assistant wrote of her: 'She says the world has used her ill . . . these men of learning have sucked her brains, and made a great deal of publishing works, of which she furnished the contents.' She did eventually become famous as a fossil hunter in her own lifetime, however. Charles Lyell even wrote to her asking her opinion on the formation of sea cliffs.

Mary Anning's discoveries inspired other scientists to work on dinosaurs. They have always been a fascination, and thousands of different species have been identified, from all over the world. The fossil record reveals that, 66 million years ago, three-quarters of the plant and animal species on Earth became extinct. In 1980, it was proposed by Luis Alvarez and his son Walter that a giant meteorite had hit the Earth. There is evidence that the site of the impact was the Chicxulub crater off the Yucatán Peninsula.

Alvarez was a Nobel Prize-winning physicist who worked

on the US atomic bomb during the Second World War, flying in formation with the plane that dropped the bomb on Hiroshima so he could measure its impact. He also played a major role in developing radar technology, including ground control systems to help planes land in bad weather. Before coming up with his asteroid theory, he scanned the Great Pyramid of Giza using cosmic rays to look for hidden chambers, and even contributed to an analysis of the Kennedy assassination film.

In March 2020, a group of scientists endorsed the asteroid hypothesis, having assessed 20 years of scientific evidence. Yet again, an example of how science works. A complex problem with a lot of supporting evidence coming to a conclusion around which a consensus is reached. The consensus in this case is that an asteroid 10–15 kilometres in diameter crashed into the jungles of the Yucatán, releasing the same energy as 100 million megatons of TNT, which is over a billion times the energy of the atom bomb dropped on Hiroshima. It left a crater 150 kilometres wide. Some of the evidence is the thin layer of iridium that has existed all over the Earth from that time – a metal which is 500 times more common in meteorites than in the Earth's crust.

I vividly remember hearing this evidence in a zoology lecture when I was a science student. The lecturer, Frank Jeal, was our favourite. He told great stories and drew fantastic diagrams of dinosaurs on the blackboard. That morning, it was raining heavily, and many of us were sitting in the lecture theatre soaked to the skin. He said: 'Sometimes you have to wonder whether it was worth us evolving to live on land.' And then he told us about the evidence for how the dinosaurs became extinct. It was spellbinding.

Without this mass extinction event, we as a species would

not have evolved to fill the niches left open by all the species that died. Birds are the descendants of the dinosaurs from all those millions of years ago. And we are descended from a small shrew-like creature called a plesiadapiform (easy to remember: 'Please Adapt and Form Me'), that survived and eventually evolved into us.

Living on the Crust

Even though lots of work had been done by fossil hunters, and by geologists dating the rocks of the Earth and the fossils within them, there was still no real sense of how the continents had formed. That changed in 1912, when Alfred Wegener came up with the theory of continental drift, defined as the slow (very slow . . .) movement of the continents across the surface of the Earth. He too fell foul of religion – especially with his father, who was a theologian.

Born in 1880 in Berlin, Wegener obtained a PhD in astronomy and later worked as a meteorologist. On a balloon trip to carry out meteorology experiments he set a new record for continuous balloon flight, at 52.4 hours (it doesn't seem he took any animals with him, and it's unclear how he went to the toilet).

He constructed the first meteorological station in Greenland and began recording the climate in the Arctic. He fought in the First World War, and was wounded twice, leading him to be reassigned to the army weather service – no doubt much to his relief. There he found time to complete a treatise on continental drift entitled *The Origin of Continents and Oceans*. He became an expert on tornadoes. His inspiration for continental drift came when he looked at the shape of the coastline of West Africa and the coastline of Brazil and concluded that they must have once been joined and then drifted apart. Wegener didn't provide a mechanism for this drift. He thought that it was driven by the centrifugal force of the Earth's rotation. He died on an expedition to Greenland in 1930 while trying to bring food supplies to colleagues. He was buried in a pyramid-shaped mausoleum in the ice.

Many geologists at the time rejected Wegener's ideas, particularly in the US, and notably at a major conference in New York, where everyone disagreed, with the exception of the chairperson of the session he was speaking in. Which shows you how science can be a very human activity. New upstart with an idea that goes against decades of work by now-old men (who are inclined to attend conferences and spout on) being denied by the mob. We'll see more examples of that in later chapters.

But the evidence began to grow and led to what philosopher Thomas Kuhn described as a 'paradigm shift'. This is how a scientific advance leads to a fundamental change in our understanding of something. It usually happens not because people become convinced, but because proponents of the old explanation die out, and those of the new idea increase in number.

Earlier studies on mountain formation provided evidence – one example being the work of Scottish geologist May Ogilvie Gordon, who had speculated that the Dolomite mountain range in the Alps was formed by 'crust-torsion', where the Earth's crust had been twisted and folded into shape, by two land masses hitting off each other. She won the Lyell Medal from the Geological Society of London for her work. Another key piece of evidence came from the observation that rocks of different ages have a different magnetic field direction because the magnetic north and south poles reverse from time to time. In the 1950s, studies on the ocean floor described mid-ocean ridges and the sea floor spreading out from them pushing the continents apart. When these ridges were examined, scientists could see that the magnetic field was alternating, moving away from the centre of the ridge. 'Stripes' of magnetic field direction could therefore be seen in the rocks spreading out on both sides of the ridges. These bands of rocks were effectively on a conveyor belt, being pushed outwards.

The idea of plate tectonics was then proposed, with the Earth being made up of plates. Think of these like big flat pieces of butter slipping around a heated frying pan as the bottoms melt. The plates that form the solid rocks of the continents are sitting on heated magma in the Earth's crust, and move around. Mountains rise when plates crash into each other, and earthquakes occur when there is a fracture between plates.

The inner secrets of the Earth then began to reveal themselves, providing further support for plate tectonics. Inge Lehmann, a Danish geophysicist, discovered that the Earth had a solid inner core, surrounded by a molten outer core. Born in 1888, she had received her education at a school that treated boys and girls equally, which was practically unheard of at that time. She then attended Copenhagen University and the University of Cambridge, mainly studying mathematics. After graduating, she worked as a seismologist in Copenhagen, setting up seismological observatories in Denmark and Greenland.

Her work on the Earth's core stemmed from observations of what are called step changes or discontinuities in seismic waves. One of these, observed at depths between 190 and

250 kilometres, was named the Lehmann discontinuity. The discontinuities revealed that the inner Earth wasn't homogeneous as was thought at the time, but rather composed of layers. In 1995, the American Geophysical Union established the Inge Lehmann Medal in her honour for 'outstanding contributions to the understanding of the structure, composition and dynamics of the Earth's mantle and core'.

We now know in great detail the composition of each of the inner regions of the Earth. The inner core is a solid sphere with a radius of about 1,220 kilometres. It is believed to be made of an iron-nickel alloy, although no one has ever sampled it – don't believe those *Journey to the Centre of the Earth* movies. The temperature gets as high as 5,430 degrees Celsius, but the intense pressure turns what would normally be a liquid at that temperature into a solid.

The convection in the molten outer core generates the Earth's magnetic field. This is a bit like a natural electrical generator, where the movement of electrically conducting iron in the fluid in the outer core causes an electric current which generates a magnetic field. The outer core is a layer of fluid 2,300 kilometres thick made mainly of molten iron and nickel as separate elements. It gets as hot as 4,230 degrees Celsius. The mantle occurs outside the outer core and is a layer of silicate-based rocks about 2,900 kilometres thick. The final layer is the crust, on which we live.

John Lennon's Drumlin

As well as looking inside the Earth, there was a lot of work going on into the geology of the Earth's surface. Mountains and valleys and many physical features can be formed by erosion of various kinds – water and wind being two major influences. Ice is another striking driver of physical features. This was first proposed in 1742, when Pierre Martel, a Swiss geographer, reported how the people of Chamonix in the Alps had told him that huge boulders (now known as erratics) had been put there by glaciers – huge masses of ice that at one time covered the Alps. Erratics are rocks that differ from the rocks native to the area in which they rest, and they were always a puzzle. Might a mythical Irish giant like Finn McCool have thrown them from one place to another as he had been reputed to do with the Isle of Man – which was formerly where Lough Neagh now is in Northern Ireland? No, of course he didn't.

In 1824, geologist Jens Esmark proposed that there had been a series of worldwide ice ages. He wondered if they had arisen because of differences in the Earth's orbit around the sun. Perhaps there were periods when the orbit was a bit wider and so at times the Earth was further from the sun. Other geologists jumped on the ice age bandwagon and began speculating on how at different times the polar ice cap might have extended well down into temperate zones. The German botanist Karl Friedrich Schimper proposed the term 'ice age' and said there must have been periods of *Verödungszeiten*, which cheerily enough translates as 'obliteration'. Schimper and his friend Louis Agassiz are considered the founding scientists in the field of glaciology, defined as the study of ice and natural phenomena that involve ice.

Agassiz was born in the village of Môtier in Switzerland, the son of a sixth-generation Protestant clergyman. The pressure was on Louis, but his mother encouraged his scientific interests. He studied in Paris under Alexander von Humboldt, an eminent geographer and meteorologist. Agassiz's wife,

Cécile Braun, collaborated with him, particularly in his work on fossil and freshwater fish – his primary area of scientific interest.

In 1837, Agassiz described how ancient glaciers had flowed out of the Alps, and that even larger glaciers had covered Europe, Asia and North America, blanketing most of the Northern Hemisphere. He said that great sheets of ice had once covered Europe, just as they currently covered Greenland. He was mainly derided for these ideas. When he presented them to the Swiss Society for Natural Research at Neuchâtel, the audience response was highly critical.

But an English naturalist, William Buckland, supported him and accompanied him on a field trip to Scotland in 1840, where they amassed evidence that glaciation had happened there too. The overall evidence for glaciation involved erratics, but also moraines – accumulations of rocky debris consisting of rounded particles ranging in size from boulders to gravel and sand. Moraines are what is left behind by a moving glacier. Just like rivers, glaciers transport all kinds of debris, which then builds up. Lateral moraines form at the sides of glaciers, and terminal moraines mark where a glacier ends.

Agassiz also concluded that striations – long strips carved into rocks – were caused by the scrapings of glaciers. His conclusions finally began to be accepted, and in 1846, financed by the King of Prussia, he went to North America to study the natural history and geology there, obtaining an appointment at Harvard. He left his wife and children back in Switzerland, but after two years his wife died. He then married fellow naturalist Elizabeth Cabot Cary and moved his three children to the US to live with them.

Agassiz studied the effects of the last ice age in North

America and became one of the most well-known scientists in the world at that time. His friend the poet Henry Longfellow even wrote a poem celebrating his 50th birthday with the rather obvious title of 'The Fiftieth Birthday of Agassiz'. Agassiz named two fossil fish after the fossil hunter from Lyme Regis, Mary Anning, to give her the recognition that she was often denied.

But his legacy has been tarnished by evidence that he held racist views. When he went to the US he wrote about polygenism – or the idea that races were created separately. He wrote so extensively on the subject that he emboldened the field of scientific racism, which tries to use science to prove the superiority of some ethnic groups over others. This has led to some landmarks and institutions that were named after him being renamed, and the Swiss government has acknowledged his racism. Of course, all the better had these abhorrent views not been put into the world in the first place.

By 1875, the evidence for ice ages (glaciation) was so compelling that it was widely accepted. The first evidence is geological – rock scouring, moraines, U-shaped valleys (where a glacier has been in the usual V-shaped valley where a river flows, with the glacier carving out the U-shape), erratics and drumlins. 'Drumlin' comes from the Irish word *droimnín*, meaning 'littlest ridge'. It is an elongated hill in the shape of an inverted spoon, and is common in certain parts of Ireland. Many often occur together, giving the landscape the appearance of a basket of eggs. Clew Bay in County Mayo has a reputed 365 islands ('one for every day of the year'). They are all drumlins. One, called Dorinish, was bought by John Lennon in 1967 and became a hippy commune. Lennon purchased it for £1,700. After his death, his widow Yoko Ono sold it for £30,000, donating the money to a local orphanage. There are also swarms of drumlins in Wisconsin in the US, formed during the Wisconsin glaciation that happened from 75,000 to 11,000 years ago – the last great ice age. That was also the last great ice age to hit Europe.

The second line of evidence is chemical: the chemical composition of ice cores provides evidence for cooling, leading to glaciation. Ice cores are fascinating. They are ice columns drilled from an ice sheet in high mountain glaciers or a land-based ice sheet – for example, in Greenland. The ice forms from the incremental annual build-up of snow and can be up to 800,000 years old. The deeper the sample, the further back it was laid down, so what the climate was like at different times can be reconstructed – in much the same way sampling the lower layers of my freezer can reveal my dubious dietary choices over the last 10 years. Air bubbles trapped in the ice reveal the level of carbon dioxide. These bubbles are like tiny time capsules because they can be analysed to find out what the atmosphere was like at that particular time. Pretty cool, huh? The current record for the oldest ice core is one that was drilled from the Transantarctic Mountains. It is estimated that the ice might be up to 5 million years old. We'll come back to ice cores in the chapter on climate change.

The last piece of evidence is paleontological – about the fossils detected in the rocks. During glacial periods, cold-adapted organisms spread into wider areas, while the numbers of animals that were adapted for warm conditions declined.

As ever, science is at its best when there are multiple lines of evidence for something. There have been at least five ice ages, with the most recent (which covered most of the Northern Hemisphere) ending around 11,000 years ago.

They are all thought to have happened because of the tilting of the Earth's axis. If the Northern Hemisphere tilts a bit away from the sun, an ice age happens. This is thought to be cyclical. We are currently in what is termed an interglacial period, and the prediction is that the next ice age will begin 50,000 years from now, although how global warming might affect that is not known.

The Cow Jumped Over the Moon

Geology today continues to map the rocks of the Earth, helped greatly by satellites using advanced imaging techniques. But it also continues to make all kinds of intriguing discoveries. A recent analysis of 2.5-billion-year-old rocks from Australia (where there are a lot of old rocks) has revealed that volcanic eruptions may have spurred on the evolution of marine microorganisms that began producing oxygen – a key event in the history of life on Earth, as we will see in chapter 5.

Earth's atmosphere became rich in oxygen around 2.4 billion years ago when photosynthetic organisms evolved to capture the energy in sunlight, using it to make carbohydrates (meaning more of themselves, since the structure of plants is mainly carbohydrate). A by-product of that was oxygen. There was also what scientists call 'a short-term whiff' of oxygen 50–100 million years before it became permanent. This may have been driven by nutrients from volcanic rocks. So, volcanoes in Australia might have spurred early life that eventually led to us, as we are descended from organisms that evolved to use the oxygen that was accumulating.

New discoveries continue to be made about our Earth. In 2020, geologists reported on great blobs of rock the size of

Europe sitting on the boundary of the mantle and the Earth's inner core. These are large low-shear-velocity provinces, and have been estimated to be 100 times taller than Mount Everest.

Evidence has also emerged of a tsunami that struck land between Great Britain and the Netherlands some 8,200 years ago, submerging land there that remains submerged. This area of the North Sea is known as Doggerland. And geologists have now found evidence that some islands survived the tsunami and were home to Stone Age people for hundreds of years, before eventually submerging. This was discovered by analysing sediments from the sea floor, revealing evidence of human habitation, including bones and tools. Perhaps the Stone Age people had invented scuba diving, but formerly occupied islands seem a safer bet.

Maybe islands like these were the basis for the legend of Atlantis, referred to in the writings of Plato and said to be founded by people who were half human and half god, who created a utopian civilization with lush vegetation, exotic wildlife and a lot of gold and silver. Plato wrote that it was destroyed 9,000 years before his own time, perhaps by a volcanic explosion or a flood.

Alternatively, Atlantis could have had something to do with lost continents. Two of these have been found, under Canada and off New Zealand, but they would have existed many millions of years ago, before we humans evolved. Remnants of an ancient rainforest have also been found recently. Guess where? Antarctica. There is a layer of sediment containing ancient pollen, spores, bits of flowering plants and even root networks. The forest dates to 90 million years ago when the continent of Antarctica had a much milder climate.

Geology has now even moved beyond the Earth, to analyse what the moon and planets are made of, with many surprises along the way. The moon, much to everyone's disappointment (except the lactose-intolerant), is not made of cheese. There is effectively no atmosphere (the feeble lunar atmosphere contains less than 10^6 atoms of gas per cubic centimetre, whereas on Earth the same volume contains 10^{19} atoms), and so no erosion takes place. Also, there are no tectonic plates. It does have volcanoes, and the surface we see is the product of volcanic activity as well as impact craters from asteroids.

The Apollo missions from 1969 to 1972 brought back 382 kilograms of moon rock. My university was sent a sample to analyse, and curiously they found evidence of cow dung. Then three robotic Soviet spacecraft brought back another 326 grams, and in 2020 the Chinese robotic Chang'e 5 mission brought back 1,731 grams. From the composition of the moon rocks, which closely resemble the Earth, moon geologists are of the view that the moon formed when a Mars-sized body called Theia impacted the early Earth, creating a large ring of debris which then formed the moon. This would have happened soon after the Earth formed, some 4.5 billion years ago. Imagine the noise and the spectacle of it. Mind you, in space, no one can hear you scream, because space is a vacuum so sound can't travel through it. Sorry, sci-fi fans. Theia crashing into the Earth would have been a deafeningly silent event.

As we've just seen, the Earth itself formed around 4.5 billion years ago, about the same time as the other planets. It's thought that a disc-shaped cloud of gas and dust left over from the sun's formation gradually formed each planet by a process called accretion, whereby dust grains clump together.

By some estimates, the Earth grew by a few centimetres a year, taking millions of years to form.

The name Theia is from Greek mythology. She was the mother of Selene, the goddess of the moon. The collision with Theia helps explain how the moon formed, but also why the Earth's core is larger than expected for a body its size. Theia's core and mantle are thought to have mixed with the Earth's core and mantle.

Not only did this collision lead to the formation of the moon and build the Earth's inner core, it also resulted in a tilt in the Earth's axis, creating the seasons. When the Northern Hemisphere tilts towards the sun, it's summer there. When it tilts away, it's winter. If the collision hadn't happened, there would be no moon and no seasons. Imagine that. A much more boring place. Computer simulations have calculated that trillions of tons of material would have vaporized and melted at the point of collision, and the temperature would have reached 10,000 degrees Celsius, shining brighter than the sun. There was no life on Earth at that time, but if there had been it would have been wiped out.

Initially the moon would have been ten times closer to the Earth than it is now. That must have been some sight, with it almost filling the sky. It then would have moved outwards, finding its final orbit. The moon is locked into position with the Earth, so that only one side faces us. Moon geologists are still intrigued as to what exactly the moon is made of. In 2020, a team of researchers reported that the Lunar Reconnaissance Orbiter spacecraft had managed to provide further analysis of the moon's sub-surface and reported that it was richer in iron and titanium than previously thought. The moon may yet have secrets to reveal – which as we read about in chapter 2, should happen with the planned building of a base there.

And what about the planets? A technique called spectroscopy has revealed that there are three main types: the rocky planets, the gas giants and the ice giants. (These might seem more like gang names, or perhaps wrestlers – 'Rocky Planet vs The Gas Giant Smackdown!') The two largest planets, Jupiter and Saturn, have nearly the same chemical composition as the sun. They are mainly made of hydrogen and helium. Even though they are gas giants, the gases are compressed into liquids, so we should really call them liquid giants. There are other elements on these giants, however, that form solid cores. The rocky planets (of which Earth is one) are quite different, being composed mainly of rocks and metals. Mercury has the greatest proportion of metals. Earth, Venus and Mars are similar, with a lot of iron and nickel. Uranus and Neptune are ice giants, mainly composed of frozen gases and liquids, including water, methane and ammonia.

Will we eventually go mining on the moon or on the rocky planets? The off-planet mining industry is a staple of many sci-fi films, and with iridium currently valued at around $200,000 per kilogram and rhodium at $350,000 per kilogram – combined with the environmental impact of mining on Earth – it could soon be a reality. If it does happen, I hope we show more respect to those other planets than we have ours.

4. Better Living through Chemistry

Walter White from *Breaking Bad* is undoubtedly the world's most famous chemistry teacher (sorry, Mr Gutty). As he once said, 'Well, technically chemistry is the study of matter. But I prefer to see it as the study of change.' We humans began to change things chemically without knowing what we were doing. And from what we observed, we questioned what was going on. We take chemistry for granted now, but it was some achievement to figure out all the elements that make up all things and learn how to manipulate them and make stuff. Stuff that got more and more elaborate, useful and at times downright dangerous. Stuff that could eventually end all life on Earth if we don't watch out.

The science of chemistry starts with fire, which is a chemical reaction – the combination of oxygen in the air with carbon-containing materials such as cellulose in wood. Our ancient ancestors had no notion of this, of course, they just figured out how to make fire happen and watched things burn. The chemical reaction generates carbon dioxide and water and releases energy in the form of light and heat. The nerds are right: chemistry is lit.

Without doubt, the control of fire was critical for chemistry because it allowed for all kinds of things. Fire could be used to cook food, which increased the variety of foodstuffs that could be eaten by changing consistency, destroying toxins and making things delicious. Our ancestors probably didn't like uncooked grains, seeds or meat any more than we do. Cooking food also killed germs in the food, preventing disease.

The earliest evidence of cooking goes back 1 million years, where in a cave in South Africa called the Wonderwerk Cave there are traces of cooked grass and leaves, and also animal bones. A stew with vegetables, perhaps. Cooking makes food easier to chew and digest and so it propelled us forward nutritionally and probably allowed for the evolution of our big, hungry brains. When you next watch that cookery show, remember that it was cooking that allowed our brains to evolve, so never scoff about the importance of *The Great British Bake Off*.

There is evidence of the widespread use of fire from around 50,000 years ago. For a long time, fire was viewed as some kind of magic, because it could produce heat (very important as humans moved into colder parts of the world) and light (illuminating what had been the pitch-dark of night). Fire had other uses, too, including the extraction of metals from rocks (smelting), and melting metals to make tools or weapons. Fire was also critical in the making of pottery, in which clay is heated to change the chemical structure of silicon and aluminium oxide, the predominant minerals in suitable clays. A lot of this must have happened through trial and error and observation. And it all gave rise to the famous ages we date human history by, such as the Bronze Age (3300–1200 BCE) and the Iron Age (1200–550 BCE).

All of this is of course much too long ago for us to know who these early chemists were. But someone was the first to figure out how to use fire or smelt metals, and they taught others, and then it spread among *Homo sapiens* and became one of the defining features of our species.

Other early evidence for chemistry can be found in the

Blombos Cave in South Africa, dating from around 100,000 years ago. What would we do without South African caves to tell us about our distant past? The early Walter Whites there seem to have been processing red ochre, a pigment extracted from clay. Why those ancient humans were extracting it isn't certain. Maybe to paint the walls of the cave, or themselves? Perhaps as a medicine? The Gugadja people of Western Australia were reported to have used it to heal wounds and burns.

An ochre-processing workshop was also excavated at Blombos. Abalone shells used to store the pigment were discovered, along with stones and bones used for grinding. The people who used the cave would have been hunter-gatherers, and spear points made from stone have been found in abundance in the caves, as have carved bones and necklaces made from the shells of sea snails. We know of no other creatures that deliberately made these kinds of things. The artwork on the cave walls is especially notable. It comprises nine red lines and dates from 73,000 years ago, making it the oldest known abstract drawing on rock.

The first pure chemical to be extracted was – guess what – the metal gold. Gold, silver, copper, tin and iron can all be found 'native' – meaning not bound to other chemicals as part of rocks or minerals – the classic example being gold nuggets. The Egyptians made the first artefacts from gold, which appeared around 4000 BCE. They were also among the first to extract iron from rocks, and used it to make weapons, calling them 'daggers from heaven'. The chemistry around those early attempts at metallurgy was simple enough though – just requiring a hammer to smash up the rocks and a fire to melt the metal.

But then we get to the Greek philosopher Empedocles.

I've always wondered why those Greeks have such unpronounceable names, something I was discussing with my friends Saoirse, Niamh and Tadhg in Dublin recently. Empedocles said there were four elements: earth, fire, air and water. He was out by over 100 but it was a good attempt at the time. He said they were brought together into different forms by 'love' and broken apart by 'strife'. He practised strife on himself when he jumped into the Mount Etna volcano in Sicily, which he did because he believed in reincarnation – or as the Greeks called it, metempsychosis. He was out by 100 per cent on his probability of reincarnation.

The defining of matter reflects how scientists like order – just like teenage me making lists of different types of rocks. Empedocles was something of a hero in ancient Greece. He was apparently able to cure many diseases and could also avert epidemics. (I wonder, can we bring him back?) Another Greek philosopher, Democritus, said that everything was made of indestructible atoms. Showing that academic rivalry was alive and well in ancient Greece, the philosopher Plato was said to have hated Democritus, wishing that all his books be burned. Plato was the first to write about 'elements' as the building blocks of matter, and the term stuck to this very day.

Kerotakis

Chemistry as an experimental science can be dated back to the alchemists. The earliest was Jabir ibn Hayyan, and the word 'alchemy' comes from Arabic. Jabir is thought to have lived in the eighth century CE. He had read about the four elements but added two more: sulphur (known as 'the stone that burns') and mercury. Jabir was a prodigious writer, completing more than 3,000 treatises. For the time this was astonishing, as not a lot of new stuff was being written – especially in Europe, where monks were mostly copying the Bible. Jabir's work cites a number of other alchemists who were working in Greece and the Middle East, including those with the interesting names of Pseudo-Democritus, Mary the Jewess and Zosimos of Panopolis.

Mary lived somewhere between the first and third centuries in Alexandria. She invented various pieces of laboratory equipment, notably the *tribikos*, a three-legged glass vessel; the *kerotakis*, which is a device for distillation; and the famous bain-marie, named after her, which involves a pan placed in boiling water and is still used in cooking today. Jabir used all of these in his studies. He wrote about getting substances from organic matter (ammonia from plants) as well as a book about poisons. He recorded how to make many different types of elixirs, with all kinds of properties, many of which had medicinal properties. One of his books contains 30 different chapters on processes using mercury. Jabir was a big fan of mercury. The book on mercury was preceded by descriptions of laboratory equipment. Not unlike the scientific protocols we use today.

Jabir, like subsequent alchemists, had two main goals: to turn 'base metals' such as lead into gold (transmutation) and to create an elixir of immortality. Good to have a mission. This sounds a lot like anti-science and more in the realm of magic, and yet the processes used to study alchemy were important precursors to the science of chemistry. Inexorably the magic strand and the science strand that we read about in the first chapter would wind together more and more tightly when it came to chemistry.

We can date exactly when alchemy came to Europe with the publication of the Latin translation of the Arabic *Book of the Composition of Alchemy* in 1144, at a time when Europe was just coming out of the Dark Ages. It was translated by one Robertus Castrensis. This name might sound like someone who has had some unfortunate surgery, but it translates as Robert of Chester. Perhaps because of their surnames (step forward François Hotman, Frederick Clod and Paphnutia the Virgin), alchemists were often made fun of, with the English poet Chaucer accusing them of being thieves and liars. Pope John XXII banned the practice of transmutation, thinking it was a type of sorcery – which is perhaps a bit rich given Jesus had turned water into wine.

Things got a bit more respectable in the 1500s with the arrival of Paracelsus. He moved away from transmutation to using chemistry to make medicines, saying: 'Many have said of Alchemy, that it is for the making of gold and silver. For me such is not the aim, but to consider only what virtue and power may lie in medicine.'

Paracelsus was born in the wonderfully named village of Egg (but unfortunately, as far as we know, he never analysed the chemistry of eggs). His father educated him in botany, medicine, mineralogy, mining and philosophy. At the age of 16 he began studying medicine at the University of Basel in Switzerland, and obtained his medical degree in 1516 from the University of Ferrara in Italy.

Paracelsus took up a post as a doctor in Basel after travelling widely, but also taught at the university. Unlike other scholars who always lectured and wrote in Latin, Paracelsus insisted on lecturing in German, to allow his lectures to be accessible to everyone (or at least those who spoke German). He was highly critical of doctors at that time and burned the works of Galen and Avicenna, which were standard textbooks for medical students. We'll meet them later, in the chapter on medicine. Paracelsus told doctors: 'The patients are your textbook, the sickbed your study.' He was often compared to his contemporary Martin Luther, the German priest who led the Protestant Reformation, because both were openly defiant against the prevailing orthodoxy – Luther in religion, and Paracelsus in medicine. He introduced opium into medicine as a painkiller and also got doctors to stop putting cow dung and feathers onto wounds, saying, 'If you prevent infection, Nature will heal the wound all by itself.' He gave the element zinc its name and was also the first to isolate the gas hydrogen. He is even credited with coining the term 'chemistry' – and also 'gas' and 'alcohol'. No mean feat. His invention of the word 'alcohol' might go part way to explaining this unflattering description of Paracelsus in Basel by a friend: 'The two years I passed in his company, he spent in drink and gluttony, day and night. He could not be found sober an hour or two together.' I think I would have liked him. Paracelsus also came up with the dictum *Dosis sola venenum facit* – probably when he was sober. It means that things can both cure and kill, the difference between medicine and poison being only in the dose taken.

He eventually moved to Nuremberg, but the doctors there knew his reputation as a bon viveur and prevented him from practising. He was even banned from publishing a book on what at that time was considered the greatest of medical problems, syphilis – possibly because many doctors had caught it and didn't want to be diagnosed. His book on the 'French disease' never came out, but he did write pamphlets on various aspects of syphilis, concluding that it could be caught via contact between two people. After his death, his works were revived, and medical students have him to thank for insisting that doctors learn chemistry in order to be able to treat their patients. As with so many scientific breakthroughs, it seems rather obvious now.

The Irish Are Back

Alchemy had begun to lose its power with Paracelsus, and by the 1700s a strong distinction emerged between chemistry and alchemy, the latter being only about trying to make gold and mainly involving charlatans. Anglo-Irish chemist Robert Boyle, born in 1627, is seen as the first modern chemist. He was educated at Eton and as a young man travelled extensively, including to Florence to study the works of Galileo – who was still alive at the time, although there is no record that they met. On his return to England, he devoted himself to science, building a laboratory on his estate in Dorset. He's best known for Boyle's law, which relates the pressure of a gas to its volume. The higher the pressure, the lower the volume. His most famous book was called *The Sceptical Chymist*. In it, he appealed to chemists to do experiments, and came up with the idea that any theory in science needs to be proved experimentally before being regarded as true – taking a leaf out of Francis Bacon's book.

Boyle was probably responsible for the motto of the Royal Society, the world's oldest scientific society, which he co-founded in 1660: *Nullius in verba*, meaning 'take nobody's word'. The precursor to the Royal Society is thought to have been the Invisible College, which was a small community of scientists (although that term hadn't yet been invented) who met regularly to share ideas and encourage each other, free from the constraints of the colleges that were associated with universities in Cambridge and Oxford – whose main purpose at that time was to educate clergymen. The purpose of the Invisible College was to pursue the 'new philosophy' of what became known as science.

We badly needed to rediscover the motto of the Royal Society during the COVID-19 pandemic, not least because many did take people's word – on the alleged dangers of vaccines or the beneficial effects of drugs like hydroxychloroquine or Ivermectin, or drinking bleach, as we'll see in chapter 7. One problem with those treatments was that the 'word' was occasionally backed up with data, though data of a very dubious nature – but how were non-scientifically minded people to know that?

Boyle remains our champion against nonsense and conspiracy theories, and he was most likely drawn to science because he had no interest in religious debates that couldn't be proven one way or the other. Science was different, in that data could be used to support the conclusions being drawn, and this remains a major reason for people becoming scientists, including myself.

Like me, Boyle was born in Ireland. But unlike me, he couldn't work there. He said that Ireland was 'a barbarous country where chemical spirits were so misunderstood and chemical instruments so unprocurable that it was hard to have any . . . thoughts on it'. As the Irish for experimental science is *eolaíocht thurgnamhach*, it was also difficult to pronounce.

Boyle was in many ways the model scientist. He believed in the acquisition of knowledge as an end in itself, but he was still interested in its application for the betterment of humanity. His major contributions to chemistry (other than the aforementioned law on gases) include viewing elements as the basic constituents of matter, the devising of techniques to detect chemical ingredients, coining the term 'analysis', and important works on the nature of magnetism and electricity that would inspire future scientists. He was someone of such principle that when he was offered the hugely prestigious presidency of the Royal Society, he turned it down because he would not take the Oath of Allegiance to the King. Another possible reason for his refusal was that the Royal Society excluded women. His sister Katherine was a huge help in his work as a chemist, though she went unrecognized. He died a week after her. Before the establishment of scientific societies (which all excluded women), it's likely that men and women worked closely together – something that became less evident once men began forming learned societies.

Soon after I became an FRS, I stayed at the Royal Society when I was in London at a conference of doctors who had graduated from Trinity College Dublin and who were now working in London. There was a pleasant dinner, and because the restaurant was close to the Society, I invited five of the doctors back for a nightcap. The Society has an 'honesty bar' system (no one in attendance, but you can pay for drinks by putting money into a special box) and we managed to drink a few bottles of wine, all paid for. The next morning when I was checking out, the security man said that he had seen me come in the previous night with five companions on the CCTV camera and that the rules said only one guest was allowed. He also seemed to be annoyed that we had drunk a number of bottles of wine. He said he would have to report the matter to the President of the Society and that I should expect a stern letter. I replied by saying that I would frame the letter. There was a painting of Robert Boyle behind him as he told me off, and as I left, I pointed to it and said, 'The Irish are back!'

Dynamite

In the 1700s, chemistry really took off, with lots of elements and gases being described for the first time. The star of eighteenth-century chemistry was Antoine-Laurent de Lavoisier. In the late 1700s, scientists believed that air contained something called 'phlogiston' which was released when something was burned. In 1772, Lavoisier was studying how two elements, phosphorus and sulphur, burned. He came across a study by Scottish scientist Joseph Black describing a substance called 'fixed air', now known to be carbon dioxide (CO_2), which was incorporated in chalk as calcium carbonate ($CaCO_3$).

In 1774, Lavoisier was visited by an English scientist, Joseph Priestley, who told him about an experiment where he had heated a compound called 'red calx' – or in today's parlance, mercury oxide (HgO). Priestley told Lavoisier how the air that was produced was more flammable and easier to breathe than normal air. Lavoisier realized that what was being produced – by the simple action of heat causing the mercury to dissociate into the metal mercury and gaseous oxygen – was the same component in air that allowed metals and non-metals to burn. He also noticed that the end product was always acidic, and came up with the name 'oxygene', which is from the Greek for 'acid-forming'. Lavoisier also got wind of an experiment by another English scientist, Henry Cavendish, who had produced something he called 'flammable air', which we now call hydrogen. Lavoisier went on to show that reacting oxygen with hydrogen produces water. Heady stuff.

He also came up with ways to precisely measure the weight of chemicals and a new way to name chemicals (for example, acids should end in '-ic', e.g., sulphuric acid), coming very close to devising a periodic table of elements. He also devised the metric system – which is no mean feat since it's the main way we measure everything (though is still yet to reach the USA, for reasons quite beyond me). He did work in biology too, showing that respiration in animals is actually a form of combustion – 'respiratory gas exchange is a combustion, like that of a candle burning'. He did this by measuring the amount of carbon dioxide and heat given off by a guinea pig over a set period, and then calculating the amount of carbon that would be needed to produce that amount of carbon dioxide. The chemical burning of carbon was equivalent to respiration in a guinea pig, which could therefore be viewed as a living system that combusts carbon in a controlled way. Not a bad definition of life.

At the age of 28 Lavoisier married Marie-Anne Pierrette Paulze, who was just 13 when he proposed, which was apparently acceptable then. She would go on to help him in his laboratory work, and host parties with eminent scientists of the day to discuss ideas related to chemistry. During the French Revolution he was guillotined, along with 27 associates, accused of defrauding the state of money and adding water to tobacco (he had actually been involved in a campaign to ensure tobacco wasn't adulterated with other ingredients), which was a common practice in France at that time. His friends had pleaded with the judge so that he could continue his experiments, but the judge said: 'The Republic needs neither scholars nor chemists.' After his execution, a friend said: 'It took them only an instant to cut off his head, and one hundred years might not suffice to reproduce its like.'

He was eventually pardoned by the French government, and a century after his death a statue was erected to him in Paris. The only trouble was, because of lack of funding they put a different head on it that was going spare – that of the Marquis de Condorcet. The statue was eventually melted down.

John Dalton was born in Cumberland, England, in 1766. He started work at the age of 10 because his family were too poor to support him, then at 15 he became a teacher in a small school run by his brother. As a young man he loved puzzles and mathematical problems, contributing them to *The Ladies' Diary*, a popular women's magazine, to make some money. He was also obsessed with measuring the heights of mountains, which he did throughout the Lake District using a barometer – as air pressure drops at higher altitudes. He developed an interest in meteorology and began recording the weather in Manchester, where he had a job as a teacher of mathematics and natural philosophy (as science was still called). Over the course of 57 years, he would make 200,000 data entries. Scientists really do love data and lists.

Dalton began working on gases, and from that work he came up with the atomic theory, devising a way to measure atomic weights for chemical elements. The idea of an atom being the indivisible basic unit of what something is made of had a long history going back to the Greeks – 'atom' comes from the Greek for 'uncuttable'. But it was Dalton who provided the first evidence, by calculating the relative atomic weights of six elements: hydrogen, oxygen, nitrogen, carbon, sulphur and phosphorus. He made all of these relative to hydrogen, and was able to conclude that they were all made of different atoms with different relative weights. It was like each element had a different number of billiard balls, with hydrogen having just one. This was a radical idea at the time.

Dalton then proposed all kinds of things in chemistry, such as how atoms of different elements combine to form compounds, and how in chemical reactions atoms are combined, separated or rearranged. The unit of atomic weight is called the dalton in his honour.

He became famous internationally for this work, but in a reassuring development for all scientists he often had trouble getting his work published because some of the ideas were seen as too radical, including by the Royal Society. He sometimes published himself in defiance. He was asked to give lectures at the Royal Institution in London but was criticized for having a voice that was indistinct and 'singularly wanting in the language and power of illustration'.

When Dalton died, his body lay in state for four days in Manchester Town Hall, with more than 40,000 people paying their respects. It's pretty amazing really, that everything indeed is made of molecules that are very, very, very tiny, and we owe the first scientific evidence of that to Dalton. His work has stood the test of time, which is the case with all the scientific things I'm telling you about in this book and is science's greatest strength. Take nobody's word.

In the nineteenth century, the field of organic chemistry took off when Friedrich Wöhler synthesized urea, the main nitrogen-containing substance in urine. This disproved the theory of vitalism – the idea that organic chemicals had some kind of 'vital' property and could not be made. Wöhler's work knocked that particular idea on the head. He synthesized urea from ammonium cyanate, knowing that urea was made in mammals by the kidneys, as a waste product of digestion. He wrote to the Swedish chemist Jöns Jakob Berzelius: 'In a manner of speaking I can no longer hold my chemical water. I must tell you that I can make urea without the use of kidneys of any animal, be it man or dog.' Berzelius called this finding a 'jewel' for Wöhler's 'laurel wreath'. The work led to a large amount of organic chemistry, with all kinds of interesting chemicals being made.

In 1856, William Henry Perkin was an 18-year-old student, and he was given a challenge by his professor: to synthesize quinine (used in the treatment of malaria) from coal tar, a crude organic substance. He did these experiments in his apartment which was on the top floor of a house in London's East End. He failed to make quinine, but he noticed a by-product that was a deep purple colour. He had accidentally invented one of the first synthetic dyes, which was named Perkin's Mauve. This became all the rage when Queen Victoria and Empress Eugénie (wife of Napoleon III) wore dresses dyed with it. It became so commonplace that the fashion for it was known as the 'mauve measles'. He went on to make Perkin's Green and other dyes, setting up a factory beside a canal which was said to change colour depending on which dye was being made.

Perkin's work inspired many in the field of organic chemistry and chemists began to wonder about the structure of these compounds, and they discovered it came down to the interesting properties of carbon. German chemist August Kekulé showed in 1858 that organic chemicals had carbons all joined together. He was the first to depict organic chemicals with self-linking carbons. One organic chemical – benzene – actually contains a ring of carbons, joined like beads on a string. Kekulé was inspired by a dream he had on a bus of a snake eating its tail. He must have been on something at the time (chemists!). Carbon's ability to assemble into chains and rings and also to bind to other chemicals such as nitrogen, oxygen and hydrogen made it the ideal basis for making an almost infinite variety of molecules – much like Lego, only smaller.

Then, in 1867, the Swedish chemist Alfred Nobel manufactured the first explosive: dynamite. This is based on the carbon compound nitroglycerin. See – I told you how explosive the interest in organic chemistry had become! He made so much money from it but felt so guilty because it was being used to kill people, that he gave a lot of his money away to set up the Nobel Prizes.

In 1869, Dmitri Mendeleev came up with a way to systematically arrange and order all the elements. He was working as a professor at the Saint Petersburg Technological Institute, where his PhD thesis had the deceptively simple title *On the Combinations of Water and Alcohol.* I have also experimented with both those liquids. He then moved on to trying to classify elements based on their chemical properties and noticed patterns that led him to come up with the periodic table. By arranging elements in order of their atomic weight (as described by Dalton), Mendeleev was able to group them according to their chemical properties – a particular one being the ability of an element to form oxides. When he came up with this he said: 'I saw in a dream a table where all elements fell into place as required. Awakening, I immediately wrote it down on a piece of paper, only in one place did a correction later seem necessary.' What is it with these chemists and their dreams? Mendeleev saw there were gaps in the table and predicted that new elements would be discovered to fill in those gaps.

In his later years he became obsessed with a young woman called Anna Popova, threatening suicide if she wouldn't marry him, even though he was already married. His divorce from his wife Feozva Leshcheva came through one month after he married Popova. The scandal meant that he lost his university appointment, and he spent the rest of his days working in the Bureau of Weights and Measures, where he introduced the metric system into Russia. Legend has it he was also the person who set 40 per cent alcohol as the limit for vodka, obviously using his thesis on water and alcohol to good effect. He developed an interest in petroleum and founded the first oil refinery in Russia. In 1905, 1906 and 1907 he was nominated for the Nobel Prize in Chemistry, and would have won it but for the opposition of an old scientific enemy who he had upset years previously, Svante Arrhenius, whose opinion held great sway with the committee.

Radioactive Suppositories

Marie Skłodowska Curie followed Mendeleev's work closely, and from studying the element uranium she discovered radioactivity, a major advance in chemistry. Marie was born in 1867 in Warsaw, then part of the Russian empire, hence her interest in Mendeleev. She worked as a governess for a while in Poland and fell in love with the older brother of the children she was tutoring, Kazimierz Zorawski, who would go on to become an eminent mathematician. But his family frowned on the relationship because of Marie's poverty and refused to allow it. As an old man, Kazimierz would sit in front of the statue of Marie that had been erected in her honour in Kraków, presumably wondering about what might have been if they hadn't been forced to split.

Marie moved on to study physics, chemistry and mathematics in Paris with her sister Bronisława. They never had much money and Marie used to keep herself warm by wearing all the clothes she had. She met French physicist Pierre Curie, who helped her find laboratory space for her work. They were married in 1895 but neither wanted a religious service. Marie wore a dark blue outfit instead of a bridal gown, and this would serve her for many years as her lab coat. There can't be too many people who turn their wedding clothes into their lab coat, but such was the passion of Marie Skłodowska Curie.

Her discovery of radioactivity built on the work of Henri Becquerel, who found that uranium emitted something that resembled X-rays. X-rays had been discovered the year before by German physicist Wilhelm Roentgen, although he didn't know what they were. Curie, along with her husband Pierre, showed that uranium made the air around it conduct electricity, and she hypothesized that whatever was causing it was coming from the uranium atom and not from it interacting with something else. While studying a uranium mineral called pitchblende, she discovered another substance that was even more radioactive than uranium, naming it thorium. Further work on pitchblende, which involved her and Pierre shovelling tonnes of it into their laboratory because the amounts of uranium and thorium in it were relatively small, revealed yet more radioactive wonder in another element named radium. Their laboratory glowed in the night because of all the radioluminescence, which is the light that is produced in a material when it is bombarded with ionizing radiation.

She always remained loyal to Poland, teaching her daughters Polish and naming another radioactive element she discovered 'polonium', after her native land. In the course of their work, the Curies also coined the term 'radioactivity'. These are amazing achievements. To discover so many new elements – and radioactivity to boot – none of which were known before. An advance that boggled the minds of chemists everywhere. And all from the hard labour of shovelling tons of pitchblende wearing a cut-up wedding outfit.

Marie's discoveries led to the field of atomic physics. She found that radium could kill cancer cells, beginning the era of radiotherapy for cancer. Her fame began to spread, and in 1903 she and her husband were to visit the Royal Institution in London, but only Pierre was allowed to speak – there being a ban on women, as with the Royal Society excluding women from its membership. It was at the Royal Institution that Pierre suggested that, because radium could cause deep burns, it might be used to burn tumours. It had in fact already been tried to treat a painful skin condition called lupus vulgaris, a form of tuberculosis, and this led to radium being used against skin cancers. It proved successful against various other cancers too, including throat cancer and uterine cancer.

Paradoxically, chronic exposure to radiation can also cause cancer, since it can cause mutations in DNA which in turn can lead to tumours developing. That wasn't known at the time, and various products were sold containing radium, which was claimed to bring health benefits. Doramad Radioactive Toothpaste claimed to 'increase the defence of teeth and gums'. Vita Radium Suppositories were recommended to make 'weak and discouraged . . . men bubble over with joyous vitality'. If that didn't work, men could try the scrotal 'Radiendocrinator', which involved sheets of blotting paper soaked in a solution of radium salts, which were then placed under the scrotum with the instruction 'Wear at night. Radiate as directed.' The inventor of both the Radiendocrinator and radium tablets, William Bailey, claimed to have drunk more radium water than any other living man – he died of cancer of the bladder in 1949. No wonder my French collaborator wore his *boules testicules*.

Radium, discovered by the Curies, proved useful for the treatment of various cancers up until the 1980s, when it was replaced with other radionuclides such as cobalt that are easier to produce and more potent.

Radiotherapy wouldn't have happened without Marie Skłodowska Curie's fundamental research into radioactivity. She said as much in a 1921 speech at Vassar College in New York: 'We must not forget that when radium was discovered no one knew that it would prove useful in hospitals. The work was one of pure science. And this is a proof that scientific work must not be considered from the point of view of the direct usefulness of it. It must be done for itself, for the beauty of science, and then there is always the chance that a scientific discovery may become like the radium – a benefit for humanity.' The battle to continue funding basic, so-called blue-skies research continues today. Marie went on to win two Nobel Prizes, the only woman to have achieved that so far. She shared the 1903 Nobel Prize in Physics with Pierre.

In 1906, Pierre was killed in a traffic accident in Paris, and Marie was devastated. Scandal broke in 1911 when it was revealed that Curie had been having an affair with physicist Paul Langevin, a former student of her husband's, who was married. This led to the Nobel Prize committee attempting to prevent her from attending the ceremony in 1911 where she was to be awarded a second Nobel Prize, citing her questionable morals. Again, Svante Arrhenius led the criticism (what is it with that guy?). To which Marie replied: 'There is no relation between my scientific work and the facts of my private life.' This led to an ugly incident, fuelled by sexism and xenophobia – she was accused of being a 'foreign Jewish homewrecker' and an angry mob protested outside her house in Paris. That year she was also turned down for membership of the French Academy of Sciences, losing out by one vote to Édouard Branly, who had helped Marconi develop the wireless telegraph. But who's ever heard of him?

Apart from her pioneering work on radium as a treatment for cancer, she also got involved in helping soldiers in the First World War. She saw the need for X-rays in the battlefield to help with surgery on wounded soldiers and established over 200 radiological units (known as Little Curies). She also attempted to donate her gold Nobel Prize medals to the war effort, but the French National Bank refused them. She said, 'I am going to give up the little gold I possess. I shall add to this the scientific medals, which are quite useless to me.' She also donated her prize money.

In 1921, Marie toured the US raising money for research into radium. She was honoured at the White House by President Warren Harding, who presented her with one gram of radium for her to use against cancer. The French government, who had never honoured her, became embarrassed and offered her the Legion of Honour. She turned it down, most likely because she didn't believe in awards for science. She and her husband often refused awards and medals, and any monetary gifts or awards she received she gave to the scientific institutions she was part of.

Her daughter Irène Curie followed her mother into research on radioactivity, and won the Nobel Prize for work on the neutron with her husband Frédéric Joliot, making them the only other married couple to jointly win a Nobel Prize. Some family. Marie Curie died at the age of 66 on 4 July 1934, from aplastic anaemia, which may have been a consequence of her exposure to radiation during her research. She effectively gave her life for her discoveries. Her papers from the 1890s are still too dangerous to handle because of the level of radioactivity. Even her cookbooks remain radioactive. Both Marie and Pierre are buried in lead-lined coffins.

Just Because You're Smart It Doesn't Mean You're Always *Right*

The Curies' work was extremely important for the field of quantum physics, which revealed some of the inner workings of atoms. Electrons were first observed in 1897 by English scientist J. J. Thomson, who was working with a cathode tube – a device that could generate streams of electricity that fluoresced. Thomson could divert the stream with electric and magnetic fields and therefore concluded that it was made of particles which he named 'corpuscles', now known as electrons. The word 'electron' was coined by Irish physicist George Stoney. The proton was discovered in 1920 by New Zealand physicist Ernest Rutherford, who was working with J. J. Thomson. He was investigating the nucleus of the hydrogen atom, and determined that it was made of a particle which he named the proton. James Chadwick gets the credit for discovering the neutron. He joined the laboratory of Rutherford, and so here we see how science often works: a group of scientists in the same place make a number of key discoveries in one area, all of them inspiring the others. Scientists are drawn to working with like-minded people trying to solve the same problem. In the case of Thomson, Rutherford and Chadwick, that problem was the structure of the atom. Chemists needed physicists to tell them what they had found, and the physicists needed the chemists to come up with all the fancy elements and isotopes. Another example of many heads is better than one, and how collaborative great science is.

Chemists had succeeded in defining the chemical nature of the matter that makes up the land, sea, sky and the natural world of plants and animals, and had moved on to use their knowledge to make all kinds of new and interesting chemicals. In 1939 American chemist Linus Pauling came up with theories about chemical bonds, writing *The Nature of the Chemical Bond* – which has been called the most influential chemistry book of the century, as guess what, it describes the nature of chemical bonds. In the 1930s he had begun working on biological molecules describing the structure of proteins. He made a mistake in the 1950s when it came to DNA, which he said was a triple helix. As we will see in the DNA chapter, he was out by one. Later in life he became a proponent of high-dose vitamin C as a cure for the common cold, which sadly turned out to be without basis. Just because you're smart it doesn't mean you're *always* right.

Chemistry is such a vast subject with relevance to almost every aspect of our lives. The advances in the past 30 years have been huge and can be seen from the various Nobel Prize winners, which is as good a way as any to spot advances that have been made – although as we have seen in this chapter, it can be open to manipulation and prejudice.

What is striking is that many of them have been for discoveries in biology where chemistry was brought to bear. Several prizes have been given for solving the structures (or shapes) of complex proteins, including the machinery that makes ATP (adenosine triphosphate), the energy currency of all living things, the enzyme complex that makes RNA from DNA (the secret of life itself), and the highly complex structure that makes proteins, called the ribosome. These are among the most complicated structures ever seen. Unusually for chemistry, as is the case with most Nobel Prizes, it was a woman, Ada Yonath, who shared that prize with Venkatraman Ramakrishnan and Thomas Steitz. We'll read more about the chemical basis for life in chapters 5 and 6.

Two women shared the Chemistry prize in 2020 – Emmanuelle Charpentier and Jennifer Doudna – for 'the development of a method for genome editing'. They had discovered a technique called CRISPR which can be used to edit genes. Doudna grew up in Hilo, Hawaii, and was fascinated by the natural beauty of the island. In this way she was driven to understand the complexities of the natural world. When she was in school, she was given a copy of James Watson's *The Double Helix*, his account of the discovery of the structure of DNA. It had a hugely inspirational effect on her.

In university she almost dropped out, because she wasn't sure science was for her, and considered switching to studying French. But her French teacher said, *non*, she should stick with science, so Doudna did that, thankfully. Charpentier was born in Juvisy-sur-Orge in France. As a student she was something of a loner. She loved to study in the old Sainte-Geneviève Library, near Notre-Dame Cathedral in Paris. She loved the pools of light at each desk in the old atmospheric library.

In 2011 she met Doudna at an American Society for Microbiology conference that was being held that year in Puerto Rico. Again, this illustrates a very important feature of science: collaboration. And we also love to travel to nice places for conferences. Face-to-face conferences are important, especially the coffee breaks and social activities. Scientists can relax and talk about their work and new ideas, and collaborative possibilities can emerge. Most of my collaborations, on which my lab depends, started in that way. We're a very social bunch, us scientists.

Doudna and Charpentier were working on similar things, and perhaps Doudna's love of French helped the collaboration. Doudna later said of the meeting: 'I loved her intensity, which was apparent from the moment I met her.' Scientific love at first sight. They would work together on CRISPR.

CRISPR stands for 'Clustered Regularly Interspaced Short Palindromic Repeats'. CRISPR ('crisper') is easier to say. It's based on a strategy used by bacteria to target the DNA of viruses to eliminate them. It's like the immune system for bacteria. But Charpentier and Doudna realized that it could be used to target any DNA in a highly specific manner. CRISPR is a bit like having a pair of super-tiny scissors to cut a piece of DNA that's harmful, which is then stitched back up again rendering it safe. Imagine you have a dirty smudge on your nice dress. CRISPR can spot the smudge, cut it out and repair the dress. Except, in the case of CRISPR, the smudge is a genetic mutation that needs to be corrected.

The main potential applications of CRISPR include correcting genetic defects, treating a wide range of diseases, and improving growth and disease resistance in crops. The first clinical trials of CRISPR are underway to treat several different diseases. It is also very useful as a research tool to study the function of different genes. In my lab we use CRISPR to delete genes in the immune system to determine how important those genes are and what they might regulate.

Meanwhile, exotic chemicals continue to be made based on carbon – including buckminsterfullerene (which looks like a soccer ball, but one made of 60 carbon atoms in a lattice) and graphene (used in electronics, but yet to find a killer application).

That ubiquitous device that we all can't survive without, the iPhone, uses 75 elements from the periodic table. Mendeleev would be pleased. Some are so-called rare earth metals. One of these, neodymium, makes the magnets that your iPhone needs to vibrate and also operate the speakers. Meanwhile, optical cables transmit the information being sent from your iPhone. These are truly remarkable. The total length of them on Earth at the moment would allow our planet to directly connect to Uranus – that's 4 billion kilometres. They allow email messages to arrive almost instantaneously. The jury's out on whether that's a positive.

And there's more. Wearable devices have been made that can measure blood glucose for diabetes. A whole new family of materials are being used to build better sensors for all kinds of things, and we're not far away from a patch made of conducting polymers that can be put on your skin to send and receive Wi-Fi signals much like an iPhone. A *Time* magazine article has recently said we might never be offline, with tattoo-like patches showing emails or information we call up with our voices.

Although industrial chemistry is critical to the manufacture of our clothes, drugs, cars and devices, it also generates much of the pollution that can affect our quality of life and environment. Lead in petrol was banned because it can cause brain damage in developing foetuses. PCBs were banned in plastics because they can cause cancer, and CFCs in coolants in refrigerators were banned because they destroy the ozone layer, which protects the Earth from damaging ultraviolet radiation that can cause cancer.

Fortunately, we now acknowledge the damaging effects of these pollutants, and in addition to reducing their use, technology is being deployed to remove them. Much of the 8 billion tons of plastic (up to 10 per cent of the total manufactured since the 1950s) has ended up in the oceans, and now devices to scoop up and filter plastics are being tested, so that the plastic can be degraded by heat or UV light. Biodegradable plastics are being developed that can be broken down by environmental bacteria and algae into simple chemicals including CO_2. The greenhouse gases including CO_2 are also being attacked using a variety of carbon-capture technologies that aim to isolate and store the CO_2 itself or convert it to more stable solid forms – for example, by absorption into crushed rocks like basalt, where it forms carbonates, or by incorporation into biomass by photosynthesis in algae or forests. Planting more trees could also have additional benefits in terms of greenhouse gases, as certain bacteria on tree bark can absorb and metabolize methane. Chemicals called metal organic frameworks that can harvest water from the air even in dry desert climates have been developed. And better batteries that are easier to recycle are being made, which will be needed to power all the electric vehicles that will become dominant.

Hydrogels are an important type of chemical currently being explored in lots of different systems. They are defined as water-loving polymers that do not dissolve in water. There are all kinds of applications for these, including breast implants, contact lenses, wound gels to help heal injuries, environmentally friendly glue, drug delivery devices, and tissue engineering. We may be heading towards a famous show from the 1970s, *The Six Million Dollar Man*, who after an accident was rebuilt with huge strength, speed and vision because of bionic implants. Might hydrogels and other new chemicals allow us to build a superhuman?

Remember, chemistry began with people wondering about what things are made of. It may eventually save us from ourselves, if it's deployed to halt global warming by coming up with ways to capture carbon, limit pollutants and more efficiently capture and store energy. Chemistry might also treat – or, even better, prevent – diseases that our lifestyles cause, or correct genetic defects to stop diseases happening in the first place. Who knows – chemistry might eventually allow us to have our cake and eat it too.

5. Living It Up

If an alien ever visits the Earth – and let's face it, we'd all love that – the thing that would obviously be of most interest would be the life forms: the most complex things on Earth. In fact, as the alien ship approaches the Earth it would detect on the surface all kinds of life moving around, not least in our cities and towns, but also in tropical forests and in the oceans – and most other places. We humans might appear like ants to the aliens looking down on us. It is life that makes our planet special, and while our study of exoplanets might change this, so far we haven't found evidence of life anywhere else.

The chemistry of life is intensely complex, easily the most complex chemistry of all, and we're still trying to understand it. After 150 years of investigation by innumerable scientists in many countries, we know a lot about the building blocks of life and how they fit together, but we are still ignorant about a lot of things. Not least how life started. Or how life deals with staying alive and not disintegrating away, because the law of entropy states that things have a natural tendency to lose order – or put another way, structure. A sandcastle always disintegrates, and unless I'm missing something at the beach, doesn't reproduce. But living things maintain their structure, until they die and then they fall apart and disintegrate. They are therefore able to defy the law of entropy and it's still not fully clear how.

Key questions when it comes to life are therefore what are

we made of and where did we come from, and how come we are even capable of wondering about all of this? As far as we know, we are the only entity on Earth that can consider our own existence. Thinking about the question of life has caused so much angst and given rise to all kinds of interesting traits like religion. And we're still no closer to an answer. But that is in the realm of philosophy, and the subject of different books to this one.

What *is* of relevance is the chemistry of life. As we saw in the previous chapter, the science of chemistry gave us an understanding of the chemical make-up of things in the world. It was only natural that we would then turn to ourselves. We've always wondered about ourselves and how other living things on Earth operate. And so, we began doing biology. It's perhaps in biology that we have the greatest things to wonder about.

Before science was invented, we came up with all kinds of answers to the big questions about our origins, often involving religion. Every culture has a 'creation' story, be it the rainbow serpent that Aboriginal Australians say shook the Earth into life, or the creation of Adam from clay in Christian theology, with Eve being made from Adam's spare rib. But then, massive spoiler alert . . . science got involved and ruined all that.

The science of biology starts with the domestication of plants and animals: the start of agriculture. Once we cracked a way to ensure a steady food supply, the human population began to grow and grow, to much higher numbers than were possible from the previous tedious, time-consuming and potentially unsuccessful hunting and gathering that had been our main way of obtaining food. The discovery of agriculture happened independently in at least eleven separate places without means of communication between them. It was our innate cleverness and ability to observe our environment that meant agriculture was an inevitable accident waiting to happen – as it did, in lots of different places.

From around 11,500 BCE, crops like barley, peas, lentils and chickpeas were grown in the Fertile Crescent, an area stretching from the eastern Mediterranean to western Asia. Meanwhile, in China, rice was being grown from around 11,500–6,200 BCE. Sheep were domesticated in Mesopotamia around 10,000 BCE, while cattle were domesticated around the same time in both modern Turkey and Pakistan. In the Andes of South America, which at the time had no contact with Europe or Asia, the potato was cultivated around 8,000 BCE, along with the domestication of llamas, alpacas and guinea pigs. Sorghum, which is in the grass family, was cultivated in the Sahel region of Africa around 5,000 years ago.

There was a whole lot of domestication going on, and it probably happened because humans harvested wild strands of crops but then realized they could grow the crops themselves using the seeds. Animal domestication was most likely by trial and error, since some animals couldn't be easily domesticated while others could. Anyone can domesticate a sheep because they're by their nature docile. But try wrangling a Tasmanian devil into submission. I was a big fan of Looney Tunes cartoons, and one featured a Tasmanian devil often getting into scrapes with Bugs Bunny, represented as a whirlwind with skin and hair flying. Any Australian trying to tame a Tasmanian devil is going to struggle. Mind you, the Australians didn't domesticate koala bears either, but then they didn't need to as they were docile anyway. Each area where domestication happened had its own range of crops and animals based on the local climate, and so we see different species being domesticated in different places.

In Jared Diamond's famous book *Guns, Germs and Steel* he makes the case for one reason why Europeans went to the Americas as opposed to the other way around: because of the early rise of agriculture in Europe after the last ice age. This led to the development of stable agriculture-based societies. It happened in Europe first, and in a more extensive way, because of the climate being conducive to agriculture across wide swathes of the continent. Europe had barley, two varieties of wheat, protein-rich pulses like lentils, as well as goats, sheep, horses and cattle. The horses and cattle provided food but also served as working animals, allowing more efficient cultivation methods to develop. The work rate of oxen is 10–20 times that of a single human. Oxen can therefore plough a field 20 times faster than a human. Buffaloes, donkeys and camels were also used for ploughing. The Americas had maize and plantains, as well as guinea pigs and llamas – which were more limited. My calculations based on power output and difficulty of training suggest you would need around 200 hefty guinea pigs to pull the same plough as one horse, and that just won't cut it, will it?

Trade between Europe and Asia began in earnest from the fourth century BCE. Textiles, silk, spices and gold and other metals were all traded. The spice trade was especially lucrative. Black pepper, cardamom, ginger, turmeric, nutmeg, cloves and cinnamon were brought to Europe from the 1400s. The flavour of the food being eaten went from bland to spicy, but only if you could afford it, or were prepared to take culinary adventures. Fast-forward to twentieth-century Ireland though, and not many spices had made it that far. I think I am of the last generation to spend their childhoods eating bland food. My mother, who like most Irish people of her generation wasn't especially adventurous when it came to cooking, served up meat and two veg every day. We did have salt and pepper, but in the 1970s things like chilli or curry powder hadn't taken off in Ireland. But then this exotic processed food product called Vesta Curry arrived on the shelves of supermarkets. Add water to a dried powder full of God knows what and hey presto, Vindaloo! At the age of twelve my palate was stimulated like never before. It's a different story now, where my own sons prefer poppadoms to potatoes.

All this trade didn't only involve goods, it also brought in ideas and new technologies. It's therefore fairly obvious why Europe became dominant. It wasn't possible for similar societies to develop in, say, Australia, because local climatic conditions meant less diversity in crops, and it was more isolated.

The discovery of agriculture is called the Neolithic Revolution, which was the first major revolution for us humans (the next one being the Industrial Revolution). We went from being hunter-gatherers to farmers. Although this could now be seen as *a bad thing*, as it meant the food we ate became less diverse and inequalities began to exist between the haves (those who had the seeds and knowledge) and have-nots (those who had to toil in the fields). Agriculture also enabled the first fixed or semi-fixed abodes, leading to villages, towns and eventually cities. And it gave some humans (those not toiling in the fields) more time to sit around and chew the fat. This meant more exchanges of ideas and socializing. Not unlike what happens at the scientific conferences I attend today. A lot of chatting and discussion happens outside the more formal presentations, often over food or drink, and from there, new ideas begin and collaborations form.

Various ideas have been put forward as to why farming persisted. My favourite is Professor Brian Hayden's 'feasting model', whereby agriculture was about showing off. Having a feast meant accumulating lots of food and having a big party, and the person throwing the party showing how powerful they were. Nothing much has changed. I'll bet when you have a dinner party you show off. 'Have another olive, they're from Tuscany.'

The Mummy Returns

To domesticate plants and animals took a lot of knowledge, and that knowledge is what we now call biology. For agricultural knowledge to become a science, things had to be written down. The Egyptians, as ever, were among the first to write things down about plants. Their whole civilization depended on the regular and predictable flooding of the Nile, which happened every year like clockwork and brought silt, creating fertile soil. Silt is a sediment like dust, consisting of particles of rocks and metals. It's made when rock is eroded by wind or water, and flowing water then transports it. It's the minerals in silt that help plants to grow. So, the glory of ancient Egypt was built on agricultural wealth. And yet again, it was the specific conditions that allowed Egyptians to thrive, rather than them being innately better than anyone else.

The Ebers Papyrus is a famous old document that describes lots of plants. It dates from 1550 BCE and was purchased at Luxor in 1873 by German Egyptologist Georg Ebers. It gives plant remedies for various ailments – the most common being cannabis, which was said to be useful for haemorrhoids and eye inflammation. 'Doctor, Doctor, I've got piles and my eyes are swollen.' 'Take three spliffs a day and come back and see me when you can sit down.'

The Egyptians, of course, also knew a lot of biology because they were very good at mummification, or embalming. A mummy is defined as a dead human or animal whose organs have been preserved. The word 'mummy' comes from the Arabic *mumiya*, meaning an embalmed corpse, and not from the mummy that was preserved by the deeply disturbed Norman Bates in the movie *Psycho*. It was thought that Egyptian mummies naturally occurred because of the dry environment in which they were entombed. The Egyptians must have noticed that mummification could occur naturally if the body was left in a dry tomb. But then, in 2014, a 5,600-year-old mummy was found to have been preserved using embalming oils made from conifer resin and aromatic plant extracts. They also covered the body with something called natron, a type of salt that helped dry the body. Natron packets were also placed inside the body. The last step was to wrap the body in strips of linen. Mummies have been a mainstay of horror movies. British politician Margaret Thatcher made a speech at a Conservative Party election rally in 2001, saying: 'I was told beforehand my arrival was unscheduled, but on my way here I passed a local cinema, and it turns out you were expecting me after all. The billboard read "The Mummy Returns".'

Egyptians had some fairly weird practices and saw the preservation of the body after death as a key step in living well in the afterlife. As with most human activities, mummification became a status symbol, with more and more elaborate tombs being built, culminating in the pyramids.

It's likely that from their efforts to mummify bodies, Egyptians knew a lot about the inner workings of the body, and how best to help slow down entropy after death. Scanning of mummies has revealed such things as a tool in the skull which was used to break apart the brain to allow it to drain out of the nose. Perhaps the origins of 'Whenever you see a hearse go by, remember that you're going to die' was Ancient Egypt: 'Your teeth fall in and your eyes pop out / Your brains come trickling down your snout'.

The Romans also knew a lot about the inner organs of the body, mainly through a practice called haruspicy, which appropriately enough sounds like something out of Harry Potter's book of potions. It is a form of divination involving the inspection of the entrails of sheep and chickens. The goal of haruspicy was to find out what the gods were thinking – say, before a battle. The liver was especially important and was examined for lumps, smoothness and shininess. It all seems strange to our modern minds. Artefacts based on haruspicy have been discovered – notably the Liver of Piacenza. This is a bronze model of a sheep's liver, made two centuries BCE and found in 1877. The names of gods are etched into its surface. The strand of magic in haruspicy was intertwined with the strand of the science of biology.

In around 300 BCE the Greeks begin classifying plants and animals. Remember how I said scientists love to classify things and make lists? Aristotle himself named some 500 animal species and dissected around 35 of them, describing their inner organs. He was the first person to spot that cows have a four-chambered stomach – or at least to write that fact down. He wrote a lot about the life cycles and behaviour of the plants and animals around him. He made many of his observations while living on the island of Lesbos, and produced an important early book on marine biology from observations he made in the Pyrrha lagoon. He listed the five main biological processes that we still recognize as central to life: metabolism, temperature regulation, information processing, embryogenesis (the formation and development of an embryo) and inheritance. But at that time he had no idea of the molecular basis for any of them – something we now know about in great detail, as we will see. Like all scholars who studied plants and animals up to the eighteenth century, Aristotle believed that all living things were arranged on a graded scale of increasing perfection, starting with plants and then up through different animals, culminating in us humans. He, like many people still do, thought it was all about us.

From the twelfth century, Aristotle's main biology books were translated into Latin, but they didn't cause much of a stir in Europe. Galileo rejected many of the ideas, thinking them wrong. But in the nineteenth century many famous biologists, such as Georges Cuvier and Louis Agassiz (who we read about in chapter 3) admired his work. Charles Darwin quoted from him in *On the Origin of Species*, since Aristotle had written about the possibility of a selection process involving random combinations of body parts. Zoologists now consider Aristotle to be the father of biology.

The Queen Stole My Trousers

Attempts by Aristotle and others to categorize living organisms into groups culminated in the eighteenth century with the Swedish biologist Carl Linnaeus, who came up with a way of classifying all species that we still use to this day. A lot of biology happened between Aristotle and Linnaeus – mainly involving describing different species, particularly in books on herbal remedies. *Bald's Leechbook* from the tenth century is a good example. It describes how leeches can be used for bloodletting, which was thought to be beneficial for a host of diseases from headaches to pneumonia. It also lists a large number of herbal remedies. Headaches could be cured with a stalk of crosswort tied to the head with a red kerchief. The plant agrimony (also known as cockleburr) could be used to treat erectile dysfunction, but only when boiled in milk. Boiling it in Welsh beer had the opposite effect, in what is possibly the first account of brewer's droop.

Linnaeus was born in Sweden in 1707 and came from a long line of peasants and priests. As a child he especially loved flowers, and family lore had it that whenever he was upset, giving him any flower immediately calmed him down. His father gave him a patch of ground in his garden where he was encouraged to grow plants. At school, Linnaeus rarely studied and often disappeared off into the countryside to look for plants. A boy clearly on a mission. In annoyance, his father took him out of school and apprenticed him to a cobbler. But his obsession with botany had been noticed by his former headmaster, who introduced him to a local doctor who had an interest in plants. By the age of 17, Linnaeus claimed to have read all the existing botanical literature, 'reading day and night'. Why would the young Linnaeus have bothered to do that? Yet again, we don't know, but this illustrates a common theme among scientists – becoming obsessed with one thing and then making it your life's work.

He went on to study botany and medicine at Uppsala University, and while he was in his second year, he began giving lectures. These proved very popular and would attract audiences of over 300, so he must have been a great speaker. He studied under Olof Celsius, uncle to the person who devised the Celsius temperature scale. In 1732, at the age of 24 – not unlike Darwin and his *Beagle* voyage – he went on an expedition to Lapland, travelling over 2,000 kilometres and gathering and observing many plants, birds and rocks. He managed to identify around 100 plants never before described. One of his friends was Peter Ardeti, and they made a pact that if either of them died, the other would finish his work. Peter drowned, and Linnaeus finished his book on the classification of fish – yet another book on fish that has sadly entered obscurity.

Following his tour of Lapland, Linnaeus defined what we call the binomial nomenclature of classifying species. The first part of a name identifies the genus to which a species belongs – 'genus' being just above 'species' in the hierarchy of classification. The second part is the species. Our binomial designation is *Homo sapiens*, so we are in the genus *Homo*, and our species is *sapiens*.

Linnaeus travelled widely in Europe but got into trouble on a trip to Hamburg. There he met with the mayor, who showed him what he claimed to be the remains of a seven-headed hydra. Linnaeus figured out that it was made from weasels and the skins of snakes. When he revealed this publicly, the mayor was enraged, and Linnaeus had to quickly leave Hamburg. Truth to power.

Linnaeus went on to become professor of medicine at Uppsala but kept lecturing on botany, the aim being to teach students about herbal remedies. He became even more famous as a teacher, telling his students to think for themselves and not trust anyone, even him. He was also notable for promoting breastfeeding by mothers as opposed to the use of wet nurses, which at that time was common practice. He speculated that children might take on the personality of the wet nurse through their milk. His interest in breast milk led to the term 'mammal' – from mammary gland – being widely adopted.

Linnaeus had several prominent students, and they became known as 'apostles of Carl Linnaeus'. Daniel Solander was one of these students. He is notable because he was one of the botanists on James Cook's voyage to Oceania in 1768. He suggested the place they first landed in Australia be named Botanist (later Botany) Bay, because there were so many exotic plants there. He was the first to document a large number of plants in Australia and New Zealand. Imagine the excitement he must have felt at seeing so many new exotic species. He put the Linnaean classification system into overdrive and was the first Swede to circumnavigate the globe. The other botanist on that voyage was Joseph Banks. When Cook returned, satirical writers had great fun with the story of the amorous exploits that Banks got up to on Tahiti, as opposed to his botanical investigations. After a dinner in which he ignored Queen Oberea in favour of a more attractive and younger woman, the queen had his trousers stolen as he slept.

Solander died at the age of 49 of a brain haemorrhage in Joseph Banks's house. Banks, who had become president of the Royal Society, didn't publish Solander's book on the botanical specimens he had collected. He considered him a lesser botanist for reasons unknown, but he was fond of Solander, later in life writing of his death: 'I can never think of it without a feeling of mortal pain.'

The age of exploration led to many thousands of new species being discovered and described by the new arrivals, although obviously the native peoples had already described many of these species. Voyages brought back samples, but often these were put into storage, and because there were so many they were left to decay. Indeed, Banks himself brought so many samples back from Captain Cook's voyages that, to this day, 250 years later, there are still uncategorized samples in London's Natural History Museum.

All these samples led the famous German scientist Alexander von Humboldt to exclaim in the early 1800s that the listing of species was all very well (and a necessary source of data), but what was needed was a hypothesis to explain where they all came from. Even today, scientists who just document things are called 'stamp collectors' and are somewhat frowned upon. Enter Charles Darwin.

I Hate Barnacles

When it comes to biology, Darwin is of course considered the daddy of them all. It's his theory that explains where all life on Earth comes from. No mean feat. He is one of the most influential figures in human history. Darwin first studied medicine at the University of Edinburgh, following in his father's footsteps. But he hated it, finding the lectures dull. He did however learn taxidermy, taking a course of 40 hour-long lectures from John Edmonstone. Forty lectures! What could he have covered to keep it interesting? How to stuff a frog, followed by how to stuff a toad, followed by etc. etc. Formerly enslaved, now freed, Edmonstone had set up a shop as a 'bird-stuffer' in Edinburgh. He was an important influence on Darwin, telling him tales of the tropical rainforests in South America.

When Darwin was in the second year of his medical studies, he joined a club interested in natural history and began studying marine invertebrates in the Firth of Forth – showing the value of joining a student society when at university. In my case I joined the rifle club. I thought it would be fun to shoot things, but quickly realized that maybe this was not the best thing to be doing. Darwin presented a paper to the club on how black spores in oyster shells were the eggs of a skate leech, which he found out through meticulous observation. He went to lectures in geology, but they bored him too. Either he had a very short attention span or the lecturers were dull, or maybe he just enjoyed stuffing things. His father became annoyed when he heard he was neglecting his medical studies, and so he sent him to Christ's College Cambridge with the aim of Darwin becoming a country parson. Yet again, he neglected his studies, and instead did a lot of riding and shooting. He was also an inveterate gambler. I once visited Christ's College and my friend Nick Gay showed me the book Darwin wrote his bets in. This was a moment – me holding a book full of Darwin's own handwriting, documenting not the wonders of evolution but how much he'd bet on how high a specific wall was.

He did however find one book on his course compelling: William Paley's *Natural Theology: or Evidences of the Existence and Attributes of the Deity*, which made the case for intelligent design in nature – basically, that things are all a bit too complicated to have come about by any other way than God. In the late 1600s and early 1700s intelligent design was seen as a means to praise God, as was understanding the laws of physics that had been discovered by Newton. Darwin also read what von Humboldt had written about the need for some grand theory to explain life. This may have set him off on his life's mission. You can't beat having a good question to answer if you're a scientist. Mine remains 'Why do inflammatory diseases happen?' and sadly we're no closer to answering that one. Darwin, however, would manage to answer his question as to where species came from, and the answer was evolution.

After graduation, Darwin wasn't sure what he wanted to do next (like many newly minted graduates), and then one of the botanists he had met at Cambridge recommended him for a self-funded place on HMS *Beagle* with Captain Robert FitzRoy. The ship was to set off on an expedition to chart the coastline of South America. Darwin's father objected to the two-year voyage, not least because he had to provide the money, but was eventually persuaded. The voyage, possibly one of the most important gap years in the history of science, ended up lasting five years. Darwin spent more time on land than he did at sea, investigating geology and natural history. Despite suffering severe seasickness, he managed to keep copious notes and gathered many samples. As we read earlier, on the first stop ashore, Darwin noted how in Cape Verde there was a band of seashells high up in the volcanic rock cliffs. FitzRoy had given him the first volume of Charles Lyell's *Principles of Geology*, and Darwin concluded that the seashells provided evidence for Lyell's theory that land slowly rose and fell over vast periods of time. In Patagonia, he found fossil bones of enormous extinct mammals. Rather unfortunately and indeed offensively he considered the people he met on Tierra del Fuego 'miserable degraded savages', but was convinced that all humans had a shared origin. High in the Andes, he saw yet more seashells.

On the Galapagos Islands he made observations on finches, mockingbirds and tortoises on different islands, finding them slightly different to each other. He returned to England a celebrity scientist, as his letters to scientists back home had been published and the many samples he had sent back had caused great interest. He spent time at his alma mater Cambridge, cataloguing his collection. Darwin had a knack for what is most important for any scientist – close observation and careful documentation.

Six months after returning, he began speculating on what he had observed, stating that 'one species does change into another'. He began writing but became ill from overwork. He would spend the rest of his life suffering from a range of ailments, including stomach problems, headaches and heart problems. These would get worse if he became stressed, which often happened if he made social visits. He began seeing his first cousin Emma Wedgwood. He thought about marrying her, and in his notebook made a list of advantages under 'Marry': 'constant companion and a friend in old age . . . better than a dog anyhow'. Under 'Not Marry' he listed 'less money for books' and 'terrible loss of time'. Romance clearly wasn't a strength for Darwin, although he did have a happy marriage with Emma. The couple had ten children, several of whom would go on to be important scientists themselves.

He published several books, including what remains the definitive book on barnacles, which is called – wait for it – *On Barnacles*. He spent eight years working on it and at one point wrote to a friend who had asked him how the book was going: 'I hate a Barnacle as no man ever did before.' But all the while he was also thinking and writing about evolution. He mentioned his theory to botanist Joseph Hooker, saying 'it is like confessing a murder'. The general view at the time was that species didn't change; they were created individually. And the vast majority believed they had been created by God. According to the Book of Genesis, plants and trees were created on the third day, creatures that live in the sea or that fly on the fifth day, and animals that live on land and finally humans on the sixth day. This reminds me of my favourite cartoon by Gary Larson – a drawing of God with his long white beard rolling what looks like lengths of clay and saying, 'These things are a cinch!' It's entitled 'God Makes the Snake'.

Darwin was dragging his heels on publishing his theory, but then Charles Lyell contacted him to say that he had read a paper by Alfred Russel Wallace that had the same idea as Darwin. It was entitled 'On the Law Which Has Regulated the Introduction of New Species'. Like Darwin, Wallace had travelled extensively, collecting many samples. He devised the theory of natural selection from observing species in Borneo. He also studied species in the Malay Archipelago, collecting 125,000 specimens with more than 83,000 species of beetles alone. Wallace wrote to Darwin asking if he had considered humans in his theory, but Darwin wrote back that he would avoid the subject because it was 'so surrounded by prejudices', adding 'I go much further than you'. Competition between scientists being alive and well.

Wallace then sent him a paper describing his findings, giving Darwin a fright. He graciously said that he would send it to any journal that Wallace might choose. Lyell and Hooker decided to present the work of Wallace and Darwin to the Linnean Society in London (named in honour of Carl Linnaeus) describing the 'Perpetuation of Varieties and Species by Natural Means of Selection'. Darwin had planned to attend but sadly his son had recently died of scarlet fever, and he was too distraught. Wallace didn't quite get the same recognition as Darwin, but Darwin always supported him, lobbying the government to provide him with a pension, which duly happened. Wallace's most successful book, *The Malay Archipelago*, was dedicated to Darwin. In 2012, there was a campaign to boost Wallace's reputation. There was criticism that when their work was presented, Darwin's was presented first and so got priority. Being first always matters in science, although in this case it shouldn't have. Darwin's life has also been studied much more extensively than Wallace's. One reason was thought to have been that Wallace was from a working-class background whereas Darwin was from landed gentry. This turns out not to be true, as Wallace was the son of a gentleman and attended public school, though perhaps Darwin's more prominent family put more attention on his work. There were also accusations of Darwin holding up the publication of Wallace's work, but these are also false.

Mind you, nothing much happened after the presentation. It's not as if the whole world went, 'Wow! The basis for life!' In fact, the president of the Linnean Society wrote that the year wasn't marked by any revolutionary discoveries. One review of the papers presented annoyed Darwin. Professor Samuel Haughton of Trinity College Dublin criticized what he called 'the speculation of Mess. Darwin and Wallace'. Darwin mentioned this criticism in his autobiography, writing that Haughton's was the only response to the paper presented (there were no others), and summarized Haughton's verdict as 'all that was new in them was false, and what was true was old'. Yet again, professional rivalry and the criticism of one scientist stinging another.

Haughton is noteworthy for two other things. In 1880, he presented a motion to the university council to admit women to the university, a radical idea at the time. This eventually happened in 1904, ahead of Oxford and Cambridge, so it took 24 years of campaigning. The second thing concerns hanging. In 1866, Haughton published a paper entitled 'On hanging, considered from a Mechanical and Physiological point of view'. In the paper he provided equations for hanging which would ensure the neck of the person being hanged would break instantly. He knew it would be difficult to convince the British government to abolish capital punishment and so he figured the next best thing would be for the hanging to be as humane as possible. This resulted in the *Official Table of Drops* being adopted, for use all over the British empire. It was a manual used to calculate how long the hangman's rope should be relative to the weight of the person being hanged. It became known as the Haughton Drop. Now there's a thing to have named after you. It was used up until 1965, when capital punishment was eventually abolished in the UK.

Finally, in 1859, Darwin published all his findings in *On the Origin of Species*, which proved hugely popular. The fame of Darwin over Wallace is now seen as mainly being due to the excellence of this book. Wallace himself wrote that 'this vast, this totally unprecedented change in public opinion has been the result of the work of one man, and was brought about in the short space of twenty years!'

The book is one long case for evolution. In it, Darwin wrote: '[as] there is a frequently recurring struggle for existence, it follows that any being, if it vary however slightly in any manner profitable to itself, under the complex and sometimes varying conditions of life, will have a better chance of surviving, and thus be *naturally selected*.' He also wrote 'endless forms most beautiful and most wondrous have been, and are being, evolved.'

A good example of selection can be seen in dog breeds. We've been breeding dogs for thousands of years but many of the breeds we see today were only bred over the past 150 years. Compare a tiny chihuahua to a great Irish wolfhound. Those differences happened because of selective breeding by us humans. Now imagine the effects of evolution (which occurs randomly but selects certain animals over others by natural selection) over millions of years. This is the reason why there are so many species on Earth with such divergence. It's therefore easy to see how you end up with a giant blue whale or a mouse – though both are mammals. It's just time and natural selection from prevailing environmental conditions that make it happen.

Look at That Semen

The theory of evolution soon became widely accepted, although how evolution worked in chemical terms was still a complete mystery. Discovering that would take an awful lot of science. It can in some ways be traced back to the invention of the microscope, which revealed a new universe of tiny creatures but also cells – the key building block of all living things. Dutch scientist Antonie van Leeuwenhoek was the most important early microscopist, and in the 1670s was the first person to see bacteria, red blood cells and spermatozoa. He couldn't resist looking at a sample of his own semen (men!), and he ruled out once and for all that it contained tiny human beings.

Two scientists, Matthias Schleiden and Theodor Schwann (I like to remember them by thinking of two German birds on ice skates – Schwanns Schleiden) came up with the idea that all life comes from cells, which can live on their own or in multicellular organisms like us. Schleiden was a lawyer, but after suffering a bout of severe depression and attempting suicide he decided to change professions. A wise move. He went back to university to study natural sciences and developed a love of botany, but instead of documenting plants he decided to study them under the microscope. In 1665, English scientist Robert Hooke had coined the term 'cell' having looked down a microscope at a piece of cork and seen lots of small compartments which looked like cells in a monastery.

In 1838 Schleiden took Hook's observation further, and with Schwann's help came up with the idea about cells being the key to life. Schleiden was also an advocate of evolution and one of the first German scientists to promote the theory.

Schwann was a medical doctor, but his obsession was looking at tissues through a microscope. He studied nerves, muscles and blood vessels, and he's one of the few people to have had a cell type named after him, the Schwann cell, which is a supporting cell in the brain. They surround neurons and provide them with nutrients and the insulating protein myelin. He had become aware of Schleiden's work and they became friends – one working on animal cells, the other plant cells. Schwann was the first person to give a detailed description of the important digestive enzyme pepsin, which breaks down proteins, and for coining the term 'metabolism' for the chemical reactions in living organisms.

The microscope revealed the inner parts of cells, and a whole new world was uncovered there too. Schleiden, who must have been envious of Schwann's other discoveries, was the first to recognize the importance of the nucleus in cells. In 1831 Scottish botanist Robert Brown was the first to describe the nucleus as a dense circular structure inside cells. Brown is also famous for describing Brownian motion, defined as the erratic random motion of microscopic particles suspended in a liquid. In a still sample of water, motes of dust move randomly but Brown didn't know why. Einstein's first scientific contribution was to explain the motion as being due to water molecules randomly striking and moving the particles.

Brown was another doctor obsessed with plants. He went on a four-year expedition to Australia in the early 1800s and did a huge amount of botanical research there, collecting over 2,000 new species. Sadly, most of them were lost when the ship transporting them back to England – HMS *Porpoise* – sank. He did however manage to describe almost 1,200 new species. In 1831 he gave a lecture at the Linnean Society – that hotbed where biological discoveries were so often first presented – where he named the cell nucleus, but he missed the true importance of the nucleus as he thought it was restricted to a subset of plant species. Schleiden came to the correct conclusion about the nucleus because of careful observation, and saw that it was involved in cell division for all cells, be they animal or plant. We now know of course that the nucleus is critical because it contains the DNA that provides the recipe to make proteins via another molecule called RNA. Microscopy was again put to good use, since Schleiden was able to observe changes in the shape of the nucleus when a cell divided, and the appearance of a new nucleus during cell division. Yet again though, this was only observational and the chemical basis for cell division was yet to be discovered.

But then, as we shall see, in 1869 the nucleus was found to contain a chemical that became known as DNA, which proved to be the key molecule that explained hereditary characteristics. DNA is the most important molecule in biology – if not everything – as it also explains how evolution works, as well as being the molecular basis of life itself. DNA is so important it gets its own chapter, so hold your questions until then.

Microscopists also saw the structures inside cells which became known as mitochondria, a key part of the cell that converts the energy in nutrients into a form that can be used, for example, when muscles exert themselves. They were named in 1898 by Carl Benda, from the Greek (as ever) *mitos* meaning 'thread' and *chondrios* meaning 'granule' – a most appropriate name because they looked like threads contained in a granule. But what they did in cells was unknown until the 1920s, when German biochemist Otto Warburg showed that they could consume oxygen (although that wasn't definitively proven until the 1940s). A lot of biochemistry on the energetics of living things was then carried out, and this really took off in 1941 when Fritz Lipmann showed that something called ATP was used as the key source of energy for most of the things that happened in cells – and indeed in life in general. Bioenergetics was born as a branch of biology to explain how cells handle energy.

Living things need energy to defy entropy, otherwise they would be like decaying sandcastles, and mitochondria play a huge part in that. The whole energy process can be traced back to sunlight. Plants evolved to absorb the energy of mass-less particles called photons in sunlight using a special pigment called chlorophyll (which is why plants are green). The energy is used in a process called photosynthesis, where plants perform what is probably the most impressive feat in all of life. They turn sunlight and the gas carbon dioxide into more of themselves – producing carbohydrates. These carbohydrates are then burned in cells to make ATP. And that happens in mitochondria. Electrons are stripped from these biochemicals – mainly in a process called the Krebs cycle, named after the German biochemist who first described it (and whose technician once gave me some of Krebs's glassware). The Cori cycle is another important cycle in the metabolism of carbohydrates, and was discovered by Gerty Cori and her husband Carl. Gerty is also the source of one of my favourite quotes about working in science: 'The unforgotten moments of my life are those rare ones which come after years of plodding work, when the veil over nature's secret seems suddenly to lift, and when what was dark and chaotic appears in a clear and beautiful light and pattern.'

This all sounds very complicated and it is, but its uncovering by people like Krebs and the Coris is a triumph for the science of biochemistry and our understanding of life. Think of it like building a house. The photons are somewhat like batteries being thrown by one builder to another, who inserts them into a power tool to help construct the house. That's how plants build themselves. But then the energy put into making the house can be recycled – during a later renovation of the house, some of its removed parts can be burned, and the heat released can be used to power a generator to provide energy that might be used to do any number of other things. That's what happens when we eat the carbohydrates in plants, or indeed when plants eat their own stored carbohydrate. It's actually all about recycling, where the energy in sunlight is stored in carbohydrates, which then get burned to retrieve and reuse the stored energy. The energy that originated in sunlight ends up in the carbohydrates and is then released by the Krebs cycle. The energy in the batteries (photons) is used to help build the house, parts of which can then be recycled to release energy, which ultimately means the building of more houses.

The Krebs cycle goes around and around, all the time releasing electrons. These electrons pass down a chain of proteins called the electron transport chain, which is in the inner membrane of the mitochondria (they have an inner and an outer membrane). As they move down the chain, our third subatomic particle gets involved, as protons get pumped from inside the mitochondria to in between the two membranes. The build-up of protons is uncomfortable for the mitochondria, and to relieve that pressure the protons are pumped back into the mitochondria through a most impressive machine called the ATP synthase. This is a large enzyme that turns around and around as it pumps the protons, and the energy is captured to make ATP. The term 'chemiosmosis' describes how the protons are pumped across the inner mitochondrial membrane, passing the energy generated to ATP via the ATP synthase.

The electrons are finally passed to oxygen, forming water, and in fact this is the key function of oxygen in living systems: it's the terminal electron acceptor. The whole process is called oxidative phosphorylation, because oxygen is used up and ATP is formed by phosphorylation.

One scientist deserves special credit for the science behind how ATP is made. Peter Mitchell first proposed it all, and he won the Nobel Prize for this discovery in 1978. Peter obtained a PhD in Biochemistry in 1951 at Cambridge, but for work on penicillin, the antibiotic we'll read about in chapter 7. He then moved to Edinburgh and began working on ATP production, but left because of a falling-out with colleagues. He was fortunate enough to be supported by the Wimpey company – a wealthy building firm owned by his uncle, who set him up in his own research institute in Bodmin, a small Cornish town close to the moors. With his collaborator Jennifer Moyle he worked on chemiosmosis – work that would win him the Nobel Prize in Chemistry – not in an ancient academic institution, but in beautiful Cornwall.

Many Nobel Prizes have been given since for the unravelling of this overall process.

A good analogy for how this all works is to compare electrons to water in a stream. If that stream flows down over a water wheel, the wheel starts to turn, and that can provide energy for turning a mill to grind corn. In cells, the electrons that originate in the carbohydrates from plants are stripped from the carbohydrates during the Krebs cycle, cascade down the electron transport chain, and the energy is captured by protons turning the ATP synthase wheel and used to make ATP, the source of energy for the processes in life.

There in a single paragraph is how energy works in living systems, and why we need oxygen. Simple enough, right?! It is impossible not to be completely blown away by the wonder of all this. We know how it works in detail, and it involves photons and two subatomic particles – electrons and protons. As a student of biochemistry, it took me several goes to understand what was going on given all the moving parts. DNA seemed much simpler: two strands forming a double helix which separate and make copies of themselves was easy by comparison. But photosynthesis, the Krebs cycle, the electron transport chain and oxidative phosphorylation were horses of a different colour. Or horses eating hay to make ATP to get their muscles to work as they run. Credit due to the thousands of scientists who discovered it all. They didn't need to bother, did they? But their curiosity and perhaps their inspirational professor drove them to figure it out. Photons to electrons to protons, all in a living system and all about generating ATP, the energy currency of all life.

Bioenergetics and mitochondria continue to be a very active area of research. My own lab is currently working on the Krebs cycle in immune cells, as it turns out it can go a bit rogue at times in diseases like arthritis, generating toxic fumes – a bit like a car engine that overheats. Or splinters of wood coming off the water wheel and harming the miller. We work on the splinters succinate and fumarate, which are part of the Krebs cycle but for unknown reasons get over-produced in the inflammatory macrophage (the front-line cell of the immune system) and might drive inflammatory diseases. We have a lot more to learn about mitochondria in health and disease.

But equally importantly, having grappled with a particularly knotty bit of science, you know enough to get this joke: A man walks into a bar and asks for a glass of energy. The bartender hands it over and says, 'That'll be 80p.' Say it out loud if you don't get it!

Without Errors There Would Be No Evolution

Every cell in every organism on Earth uses ATP in the same way, which means one thing: all life is descended from the first cell that could do all this and that could make copies of itself. And then natural selection, as discovered by Darwin and Wallace, worked on that cell and we got the evolution of different species.

This realization led to one unrelenting question that wouldn't go away and still hasn't: how exactly did life start? As we saw in chapter 4, in the nineteenth century a chemist called Friedrich Wöhler made urea in a lab. This was a shock as it was thought that organic chemicals like urea had something special about them (vitalism). But Wöhler had shown this not to be the case. He worked in a lab built for him by his father in his house. A bit like how Mr Wimpey built Peter Mitchell's bioenergetics lab. But he eventually graduated from the University of Dad and went on to work at the University of Berlin. He was also the first to isolate aluminium and discovered the element beryllium. And his discovery with urea led to the increasing realization that, despite our huge egos, we aren't fundamentally that different to other living creatures.

The first significant experiment to address how life started was done by two scientists, Stanley Miller and Harold Urey, in 1953. They mixed chemicals that the fossil record said existed in the early days of the Earth – namely methane, hydrogen, ammonia and water – provided energy in the form of a Bunsen burner, as well as electrical sparks from an electrode, and let the whole thing run. After a few days a tiny creature crawled out . . .

I'm afraid not. But they *had* made amino acids, the basic building blocks of proteins, and from very simple ingredients that would have been on early Earth. Urey was one of the first scientists to analyse the moon rocks brought back by the Apollo 11 mission that we read about earlier. Sadly, he found no evidence for extraterrestrial life, but I bet he looked hard for it. One story claimed he found evidence for cow dung in a moon rock sample. It was said that this proved that the cow had indeed jumped over the moon. Sadly, this was not true. Conditions on the moon aren't conducive for life to evolve – it has almost no gaseous atmosphere (though surely a few cows could change that), a lack of free liquid water, 200 times the radiation levels on Earth, extreme temperature fluctuations and very few good coffee shops. Urey published many scientific papers, well into old age. When he was asked about why he still worked so hard he said: 'Well you know, I'm not on tenure anymore.'

Since then, other scientists have mixed chemicals that

were on Earth before life – notably cyanide of all things, which is poisonous to life because it blocks electron transport in mitochondria, acting as a rod to stop the waterwheel turning. In addition to amino acids, the synthesis of the key building blocks of all living things was observed, including nucleic acids (which make DNA), fats (which make the membranes that contain the contents of the cell) and carbohydrates (that are an energy source). And all from simple ingredients. It's therefore possible to use quite simple chemistry to make biochemicals – these more than likely formed the first cell over time, and set the whole thing off. At least, that's the theory. All you need for life therefore is the correct chemicals, the right temperature conditions and an awful lot of cooking time.

On Earth it took a few hundred million years of cooking for the first cell to appear following the formation of the planet. The fossil record tells us this. But once that first cell arose and could copy itself, the greatest success in chemistry happened – the evolution of life, which in essence is just a highly complex chemical process.

So, at last we had an explanation for what Darwin and Wallace had described. Evolution happens because every time a cell copies its DNA errors are made, sometimes randomly because the enzymes involved aren't perfect and make mistakes (don't we all), and sometimes driven by radiation. This gives rise to offspring with slightly different characteristics, and then if conditions change, the fittest cell that has been randomly generated will survive and have offspring that will dominate. It's a random process, not directed towards anything in particular, least of all us. The famous evolutionary biologist Stephen Jay Gould has said that if you played the tape of life again from the start you would end up with a different outcome, because of the randomness of the conditions that drive evolution. Maybe the asteroid that wiped out the dinosaurs doesn't strike the Earth. Maybe the chemical composition of the oceans is slightly different. Who knows? But evolution works, and is a feature of all life. Unless you clone something (meaning that its DNA is identical to the parent), there will be differences between different offspring and the fittest will survive given the prevailing conditions.

Your Research Is Crap

As we have seen, another key event in the story of life is the evolution of photosynthesis. Oxygen is a by-product of that process. Paradoxically, oxygen is a rather toxic molecule. When plants arose that made oxygen, guess what happened? Mass extinctions. Think of oxygen rusting everything, like it does to iron. But then evolution did its thing and some bacteria-like cells evolved to use the oxygen in oxidative phosphorylation, to get more energy from food. What happened next is striking – and yet another random event that couldn't be predicted.

One of those cells that could use oxygen crawled inside another cell that couldn't use oxygen to form a special relationship – the original double act. In 1967, Lynn Margulis was one of the first to describe this, calling it 'endosymbiosis', and we humans – and every animal known today – are descended from that first endosymbiont. The descendants of that cell that crawled inside are what we now call mitochondria. Mitochondria have their own DNA that looks a lot like bacterial DNA, since they are likely to be descended from bacteria that evolved to handle oxygen and crawled inside another cell.

Margulis is ranked up there with Darwin by many biologists, but this isn't universal; she was shunned by other scientists at the time, with one of her grant applications for research funding being rejected with the comment 'Your research is crap. Don't ever bother to apply again.' But she prevailed. She won one of biology's greatest prizes in 2008, the Darwin–Wallace Medal. She was married and divorced twice, her first husband being the astronomer and well-known science communicator Carl Sagan. Or as she put it: 'I quit my job as a wife twice. It's not humanly possible to be a good wife, a good mother and a first-class scientist – something had to go.' Things have improved, but there's still work to be done.

Those endosymbiont cells nailed it because they could use oxygen to get more bang for their buck from nutrients, and so they began to dominate. The more oxygen there is, the more intense the burning of things made from carbon. As we saw above, oxygen is the final electron acceptor as the electrons flow down the waterfall of the electron transport chain. More oxygen pushes more electrons along and the stream flows faster, turning the ATP synthase wheel faster to generate more ATP. And so oxygen helps cells burn carbohydrates and generate lots of ATP.

The endosymbiont cells became very effective battery-makers, providing lots of energy. After a while, cells formed colonies and the first multicellular organism arose, because there's strength in numbers. A gang can be a more effective unit than a loner. And then natural selection drove evolution, and we got all the species we see on Earth today – including, wait for it, us. Yet again, all that was needed was chemistry and time – a lot of time – to get to you and me. And of course, dumb luck.

Tiny Hands

There have always been detractors when it comes to the science of life, most notably among people who refuse to believe in evolution and the history of life on Earth, despite overwhelming lines of evidence. That of course is their prerogative. Creationists believe that life originated with a supernatural act of divine creation. Many believe that the Earth is a few thousand years old, or that all of life can be explained by intelligent design – meaning that there was an intelligence behind all of this rather than the random process of natural selection. But I'll stick with the science.

In the last 30 years there has been an explosion of knowledge about living systems, as we will see in the next two chapters. Perhaps the most striking fundamental question now is whether we can bring back long-lost species, or even build a wholly synthetic life form, using what we've learned from all this biology. That would mean that we truly understood biology, as to build something means you must understand it. Although, on that basis, my qualifications as a reliable narrator might be brought into question based on my failure to assemble an IKEA chair.

Life will continue to reveal its secrets – and who knows, might eventually reveal itself on a distant exoplanet. When that discovery is made it won't be especially surprising given what we know about how life evolved on Earth. There are likely to be billions upon billions of test tubes out there in the form of planets with conditions conducive to life, bubbling away over billions and billions of years, so life is bound to have evolved somewhere else. A self-replicating system that defies entropy and can evolve. And yet again, science will remove us from the centre of things, as it has always done.

Of course, it might look different to life here, given the huge number of random events that happened to get to us. For example, without the extinction event that finished off the dinosaurs the dominance of mammals that led to humans may not have occurred – and I would be typing this with my tiny T. rex hands while roaring incoherently.

6. Message in a Bottle

What was the great piece of technology that enabled us to get to the very essence of who we are? Dirty bandages. From a seemingly unpromising starting point, we figured out what it is in sperm that mixes with something in an egg and then goes on to make a human being. The advances in chemistry and biology that we read about in the last two chapters led inexorably to what many consider to be the greatest scientific advance of all: understanding the basis for the information for life on Earth, and how that information passes from one generation to the next.

The story begins in Tübingen, Germany, of all places, and with those dirty bandages. Friedrich Miescher was born on 13 August 1844. His father and uncle were both well-known physicians and professors of anatomy and physiology at the University of Basel, so he probably never stood a chance – he was going to become a doctor or scientist whether he liked it or not (a perfect example of heredity if ever there was one).

He studied medicine in Basel (imagine having your father and uncle both teaching you) but wasn't especially interested in medicine as a career. But his father insisted he gain 'practical competence' and so he specialized as an otologist, an expert on hearing. Miescher himself had poor hearing because of an infection in childhood. I'm imagining him saying 'Can you hear me?' to a patient, them shouting their reply, and him writing down: 'This person clearly has hearing difficulties because they haven't answered me.' How did Monty Python miss that one?

Thankfully for everyone concerned, Miescher decided to pursue a career in research instead. In 1868 he went to study further in Tübingen, most likely under the influence of his uncle Wilhelm His, who was of the view that the 'last remaining questions concerning the development of tissues could only be solved on the basis of chemistry'. This was a prescient thing to say in the late 1800s, as chemistry would indeed be very useful in biology in the coming decades.

In Tübingen, Miescher went to work in the laboratories of Felix Hoppe-Seyler, who was a pioneer in the field of biochemistry – or as it was called then, physiological chemistry. Chemists were using the chemical techniques they had invented for other purposes to study the chemistry of living organisms, or more commonly tissues or fluids taken from them. Hoppe-Seyler himself was one of the first to study proteins, specifically haemoglobin, the protein that carries oxygen around our bodies. He had coined the term 'proteid', meaning 'primary', because things like haemoglobin were known to have a primary role in our bodies. The word 'protein' was already in use, however, and so 'proteid' never caught on.

At that time, Miescher was Hoppe-Seyler's only student. But he wasn't especially interested in proteins. He wanted to know what cells were made of, and he began by trying to prepare cells that were abundant in the glands in our bodies we now call lymph nodes. He knew these glands were chock-full of cells (which would eventually become known as lymphocytes). The only problem was he couldn't purify enough of them.

Hoppe-Seyler told him to stop working on glands and to move on to something that must have seemed unprepossessing but would prove very useful – pus. Hoppe-Seyler knew that bandages (a recent invention) soaked up pus, and it was known that pus teemed with cells because it had been examined with a microscope. Most of the cells in pus are dead, but it is still a great source of material. As we know now, pus is full of dead neutrophils, an important immune cell type that can kill bacteria. Pus is the result of the battle waged between neutrophils and bacteria. They are why pus is a yellowy-white colour, because they themselves are a type of white blood cell.

Presumably because Hoppe-Seyler was his supervisor, Miescher began by looking for proteins. He found that pus was full of proteins, and he tried classifying them. This proved too difficult, because methods of separating proteins from each other were still primitive. But then Miescher noticed something. When acid was added to pus, something precipitated – a white substance appeared. And then, when he added alkali, the precipitate disappeared. He found that if he used nuclei from the cells they were especially rich in this stuff that precipitated and then dissolved.

Miescher even came up with a way to carefully separate nuclei from the rest of the contents of the cells. But what might this substance that appeared and disappeared be? He first mentioned it in a letter to good old Uncle Wilhelm: 'In my experiments with low alkaline liquids, precipitates formed in the solutions after neutralisation that could not be dissolved in water, acetic acid, highly diluted hydrochloric acid or in a salt solution, and therefore do not belong to any known type of protein.' Ah! Discovery! And a puzzle to solve, which scientists love. Because it was from the nucleus, he named the precipitate 'nuclein'.

This letter from Miescher to his uncle, written in 1869, is

the first description of what we now know to be – wait for it – DNA. The molecule that in so many ways ultimately explains all you need to know about life. It's the basis of heredity, the passing on of physical or mental traits from one generation to another. It's the recipe to make proteins (the rule being DNA makes RNA makes protein – more on that later), which do all the heavy lifting of life. Random changes in DNA are the basis for evolution. No DNA, no life. No DNA mutation, no us. And no forensics (DNA fingerprinting to identify criminals), not much in the way of diagnostics (the PCR test for COVID-19 depends on DNA, as do many others), no looking back into our own past (using the DNA analysis of archaeological samples), no massive advances in biomedicine (since we use the sequence of DNA to understand many diseases, including cancer) – and who knows what else? Maybe computers based on DNA. Or a machine/human hybrid controlled by DNA that we will call the Borg.

This all began with a pus-soaked bandage and a chemical that precipitated in acid and dissolved in alkali. Modest beginnings. Miescher had no idea of the importance of what he had found. And there's a good chance we wouldn't have known about his discovery either, except for the fact that Uncle Wilhelm compiled a two-volume collection of his nephew's work – including that letter he sent. Miescher only published nine scientific papers in his entire life, although that probably wasn't too bad an output in those days. He wouldn't have survived in the vicious world of academia now, where quantity of publications trumps quality and those who shout loudest are more likely to get on. All Miescher really wanted was the satisfaction of a scientific experiment well done.

And so, he set about chemically characterizing this 'nuclein'. He could tell that, unlike proteins, it contained no sulphur. But it did have a lot of phosphorus. He wrote, 'We are dealing with an entity . . . not comparable to any hitherto known group.' He might as well have said: 'It's life, Uncle Wilhelm, but not as we know it.' He did figure out that the amount of the substance increased when a cell divided, and that intrigued him. So he wrote up his work for publication and gave it to Hoppe-Seyler to read. And what did Hoppe-Seyler do then? He was sceptical of his student's work, and so he repeated the experiments for himself. Imagine that nowadays! We live in a scientific world where supervisors like me hardly ever do laboratory work. We're too busy writing grant applications to fund the work in our labs, or setting up collaborations with other scientists. Or showing the work of our students and postdocs at conferences and getting the credit. Mind you, we call that 'networking'.

Once Hoppe-Seyler was convinced, the paper was published along with another article by another student in the lab, Pál Plósz, who had repeated Miescher's work himself using nuclei prepared from red blood cells taken from birds and snakes. Unlike our red blood cells, theirs have nuclei. These were a favourite cell type of Hoppe-Seyler, who had purified haemoglobin from them.

A different biological system leading to the same result – this made Miescher's discovery even more compelling. Hoppe-Seyler wrote an accompanying article in which he said that Miescher's work had 'enhanced our understanding of the composition of pus'. Thanks, Felix. But he did also say that the work was of 'great importance' because it was the first to study what the nucleus of the cell was made of. Miescher himself came tantalizingly close to describing what

DNA does: 'Knowledge of the relationship between the nuclear substances, proteins and their closest conversion products will gradually help lift the veil which still utterly conceals the inner processes of cell growth.'

Miescher went back to Basel and was appointed to the chair of physiology, a post previously held by his father and his uncle, who had moved to Leipzig. He'd got his dad's and uncle's old job, but he wasn't happy. He wasn't given much lab space (a common problem for newly appointed academics) and longed to be back in Tübingen. Still, he was able to show that sperm cells were a great source of nuclein, and salmon sperm was a very rich source. He figured out that nuclein was an acid, and was also able to show that it had at least four types of chemicals called bases – something that would become important, as we will see.

Miescher worked on sperm from carp, frogs, cocks and bulls. If ever there were a lesson in choosing your experimental system carefully, it's the work of Miescher. Nuclein was much easier to isolate from pus or sperm than, say, a solid organ like the liver, as breaking open the liver tissue destroyed the nuclein. We now know that the reason why pus is a great source of DNA is because neutrophils release their DNA in the form of neutrophil extracellular traps, or NETs. The neutrophil casts these NETs out to capture bacteria. DNA therefore has two jobs in neutrophils: it's the genetic material, but it can also be used to capture bacteria. Evolution often favours molecules that can do more than one thing, because they are highly efficient. A bit like how your iPhone can be a phone, camera, alarm clock, torch, etc. etc. Miescher therefore definitely chose the right material to work with, because pus is chock-full of NETs from neutrophils, and NETs are mainly made of DNA.

His work eventually attracted the attention of embryologists who were trying to understand what happens when sperm fertilizes an egg. Sperm has an interesting history in its own right. In 1677, a young medical student from Leiden, Johan Ham, brought the famous microscopist Antonie van Leeuwenhoek some pus mixed with semen to examine. What a delightful gift! Ham said it had been produced by a 'friend' who had 'lain with an unclean woman'. Nice. Ham had looked at the sample with his own microscope and been alarmed by what he saw: lots of tadpole-like creatures. He got Leeuwenhoek to check the sample, and the microscopist confirmed what Ham had seen. Leeuwenhoek then did the obvious thing. He looked at a sample of his own semen and reported on huge numbers of tiny creatures (as he saw them) thrashing away. They were named *spermatozoa*, which means 'semen animals'.

By the time of Miescher, embryologists had figured out that a sperm could fertilize an egg, which would then develop into a foetus – but the question was how? They had used a microscope to watch a sperm latch on to an egg and then seemingly release something into the egg. Miescher himself wrote: 'If one wants to assume that a single substance . . . is the specific cause of fertilization, then one should undoubtedly first and foremost consider nuclein.' But he had his doubts, as he had shown that – chemically – nuclein from the sperm of different species all looked remarkably similar, and yet each species was very different.

Miescher was burdened with giving many lectures to undergraduate students and was constantly being asked to work on surveys of nutrition, which was suitable work for a biochemist to be doing. Nutrition is all about eating food and extracting goodness from it, and biochemists were interested in the process. He began overworking. A student reported that when the lab ran out of glassware, Miescher brought in his fine porcelain dinnerware to use, and even spent his wedding day in his laboratory trying to do more experiments. It's not clear what his new wife made of that. Sadly, he contracted tuberculosis and died at the age of 51. But work on nuclein continued when another scientist in the Hoppe-Seyler lab took up the challenge.

Techniques in chemical analysis had improved, and in the late 1800s Albrecht Kossel showed that as well as containing phosphorus, nuclein indeed consisted of the four bases (a base being something that will react with an acid to form a salt) previously described by Miescher, in what we now call nucleotides, but also sugar molecules. He spotted that the sugar part was a sugar called ribose, but it was missing an oxygen – hence 'deoxy'. And so he renamed nuclein – wait for it – deoxyribonucleic acid, and we have the best-known scientific acronym ever: DNA. Kossel won the Nobel Prize in 1910 for his work on DNA. The first of many Nobel Prizes awarded for research on DNA, as we will see.

It was figured out that chromosomes, which are copied when a cell divides, contain nuclein. But for a long time after Miescher's death, nuclein received little attention. Because proteins were so much more chemically complex, everyone assumed that the secret to heredity would be found in them. The building blocks of DNA are the four bases that Miescher and then Kossel figured out were in nuclein. Meanwhile, proteins are made of 20 different biochemicals called amino acids in various combinations, and therefore have a lot more diversity.

Proteins are the workhorses of all life. Different proteins do many different things, from transporting substances around the body (like haemoglobin, which transports oxygen), to digesting food and extracting energy (enzymes), to holding your body together in your joints (collagen), to acting as key signals in your immune system (cytokines), to locking on to and helping the immune system eliminate microbes (antibodies). The list goes on and on, and biochemists in the first decades of the twentieth century kept finding out more and more interesting things about proteins. In 2022, artificial intelligence would be used by the companies Meta and Alphabet to describe the shape (structure) of 800 million different proteins from viruses, bacteria, plants, animals and fungi. DNA in chromosomes seemed more like a boring scaffold, comprising the four bases repeated ad nauseam, with the phosphorus and sugar somehow acting like rivets to hold it all together. Surely it would be in the majesty of proteins that the secret of heredity would be revealed?

What Is Life?

In 1865, around the same time as Miescher was working away in Tübingen, a Moravian monk, Gregor Mendel, had come up with the laws of heredity by working on, of all things, peas. If they had ever met for dinner, Mendel could have brought the peas, and Miescher the salmon (and hopefully not pus or semen). But what was a monk doing working on peas?

Mendel was born in 1822 in Hynčice, then Heinzendorf bei Odrau, on the Moravian-Silesian border in the Austrian empire. From a farming family, he worked as a gardener and was an avid beekeeper, but became a monk because it paid for a good education, and he said the monastic life would spare him 'the perpetual anxiety about a means of livelihood'. He was born Johann, but on becoming a monk he was given the name Gregor.

He went to the monastery in what is now Brno to train to be a priest. He also studied to be a teacher, but failed his exams and was sent to the University of Vienna to continue his education. He failed the teacher exams again, but returned to Brno and began working on variation in plants in a two-hectare experimental garden that he set up in the monastery. It looks like his farming background and learning about heredity in sheep from one of his professors, Johann Nestler, inspired him. He was also taught by Christian Doppler. (Yes, for those wondering, the originator of the Doppler effect, which happens when a police car passes you by and the sound of the siren drops a tone.)

All this education stood him in good stead, even if he had failed his exams. Beginning his experiments on peas, he studied them closely – probably more closely than anyone had ever done – and became a pea obsessive. He examined seven traits in all: seed shape, flower colour, seed coat tint, pod shape, unripe pod colour, flower location and plant height. He first concentrated on seed shape, which was either smooth or wrinkled. Over seven years he bred 28,000 pea plants, and from crossing, say, a plant grown from a smooth seed with one from a wrinkled seed, he was able to figure out the laws of inheritance. He realized traits can be passed on to subsequent generations, but some traits will dominate, and others will be what's termed recessive. However, he had no idea what the basis for this was.

His first publication on his findings was cited a total of three times over the course of 35 years. This means that, in more than three decades, only three other scientists referred to Mendel's work in their own studies. Nowadays, important papers are often cited thousands of times. Not even Charles Darwin, who was working on his theory of evolution at the time, had heard of his work. Mendel knew his research was being overlooked but was confident it would prevail. He said to a friend: 'My time will come.' Did it ever.

Like Miescher, Mendel eventually became overburdened with administrative duties and died at the age of 61. His successor burned all his papers because of a dispute between them over taxation. How petty was that? But his work wasn't forgotten. In the early 1900s, other scientists repeated his findings and rediscovered his publications, forming the basis of the science of genetics. His work is important in the story of DNA because it inspired scientists to try to find out the chemical basis for what genes are made of. They get passed to the next generation, and then confer traits on the offspring. But what exactly is making the seed of the pea smooth or wrinkled, or tall or short? The traits can be dominant or recessive, so if a tall plant is bred with a short plant, the offspring will be tall, but why is that? The tall trait is dominating over the short trait. To get a short offspring you must breed two short plants. But the unrelenting question became what are the genes that confer these traits made of, and how do they work?

It would take until 1988 – over 120 years after Mendel's work on peas – for that question to be answered. Starch-branching enzyme 1 controls the level of sugar in the seed by breaking down sugar. One form of the gene encoding the enzyme is less active, and this leads to more sugar, making the pea smooth. The more active form decreases the amount of sugar, and so the pea becomes wrinkled. During fertilization, the pea gets one copy of the gene from one parent, and another from the other parent. Let's say the one from the 'father' pea makes the really active enzyme, and the one from the 'mother' is the less active form. The gene that makes lots of active enzyme dominates, as it decreases the sugar content and the pea becomes wrinkled – even if the less active form is there. If both parents have the less active form, the pea will be smooth.

Mendel had no idea of any of this, however. He was just happy to describe how the traits were passed on. But how could stuff that Miescher had purified from pus taken from bandages have anything to do with how peas inherit traits from their parents?

This seemingly ridiculous prospect would have to wait a few more years. In 1929, a scientist called Phoebus Levene started working on DNA, with no idea what he was looking for. He took up where Kossel and Miescher had left off, and looked at the four bases in DNA that Kossel had characterized. He suggested that maybe the four bases were linked together by phosphates, and so DNA was some kind of polymer, meaning a substance built from a large number of similar units – i.e. lots and lots of the four bases, all tied together like one big necklace, with each base being a different coloured bead.

The discoveries came thick and fast. A large number of scientists began working on the issue of heredity, drawn to the challenge because of its importance for life. And a veritable menagerie of creatures was studied – adding to human pus, salmon sperm and peas. Sea urchin eggs, which are large and easy to study, were used to demonstrate that DNA is indeed in the nucleus, but that another nucleic acid called RNA – which is very similar to DNA – is outside the nucleus in what is called the cytoplasm. Nobody knew what that meant, but it would become remarkably important.

In 1927, Nikolai Koltsov suggested that whatever genes are made of they must have two strands, and that the sperm and egg have one strand each. When the sperm fertilizes the egg, it squirts one of the strands into the egg, and the two strands come together and then do something to give the organism all its traits, be they dominant or recessive. But he had no idea what those strands were made of.

In 1937, we got the first image of DNA. English scientists William Astbury and Florence Bell took a picture of a DNA crystal with an X-ray machine and saw that it was very regular. Their paper in 1938 described the structure of DNA as a 'pile of pennies'. This was a bit off-putting, again suggesting that DNA was just some kind of scaffold and couldn't possibly contain the information needed to make a pea smooth or wrinkled, or whatever the trait might be.

Erwin Schrödinger, the Nobel Prize-winning physicist, then got involved. In a series of lectures called 'What is Life?' – which he gave at Trinity College Dublin in 1943 – he speculated on what a gene might be. He had read a famous paper (called the 'green paper' because of the colour of its cover, not because it was about peas), published in 1935 by three scientists, Timofeeff, Zimmer and Delbrück, which had shown that X-rays could cause changes in fruit flies that passed on to the next generation. X-rays again, but this time causing mutation. The authors had concluded that 'a mutation is a molecular rearrangement within a particular molecule, and the gene is a union of atoms within which a mutation, in the sense of a molecular rearrangement, can occur'.

This told Schrödinger that a gene was chemical in nature. He knew it must have information and that it must be stable, and so he said a gene must be an 'aperiodic crystal'. He chose these words carefully. If something is periodic, it's pretty boring, because it's repetitive. He gave the example of the repeating pattern on wallpaper. But if something is aperiodic, it's much more interesting – and here he used the analogy of a tapestry. Life is in no way boring, he said, so whatever a gene was, it must be aperiodic (meaning not repetitive) and it must be stable – hence the word 'crystal'.

Schrödinger published the contents of his lectures in a book – *What is Life?* – that had a major influence on what came next. I was honoured to be able to celebrate the 50th and 75th anniversaries of *What is Life?* in 1993 and 2018, through a series of lectures in Dublin by some of the world's most influential scientists – who delivered their own modern take on the fundamentals of life. The respect they all had for Schrödinger's predictive work meant these lectures attracted six Nobel Prize winners to Dublin. Many reminisced on how they had read *What is Life?* and what an influence it had been on them. One of the Nobellists, Mike Rosbash, remembered how at the Nobel Prize dinner he'd had to wait an eternity at the top table during the King of Sweden's speech, dying to relieve himself. It seemed to be his main memory of the whole event.

And the same year that Schrödinger's book was first published, 1944, we get what is now seen as the 'killer experiment' that identified DNA as the molecule containing the information to govern biological traits.

Desoxyribonucleate

It was one dirty little experiment by three scientists – Oswald Avery, Colin MacLeod and Maclyn McCarty – working at the Rockefeller University in New York. They were studying a phenomenon described a few years earlier by Frederick

Griffith, who had shown that one type of bacteria can transfer a substance to another type that can change the properties of that second type. Yet again, this involved smoothness. What is it with these guys and smoothness? Griffith's work showed that the trait that makes a bacterium called *pneumococcus* smooth could be transferred to another 'rough' form and make it smooth. So smoothness could be transferred between bacteria. And the smoothness trait could even be transferred from a dead bacterium into a live one, so it had to be chemical in nature. But the question was – what was the chemical?

Oswald Avery had been working on that question since 1934. He was an unlikely person to be doing so, as he'd got the lowest score for bacteriology as an undergraduate. He'd started his research career by studying the effect of yogurt on bacteria in the human gut – an area that would become fashionable again 100 years later. Colin MacLeod joined his lab, but progress was slow, and MacLeod needed publications on other topics for his CV (that same old story), but in 1940 they returned to it. Maclyn McCarty then joined the team in 1941 and brought his biochemical skills to the table.

By 1942, they realized that whatever the stuff was, it was active at a dilution of one part per 100,000,000. Yes – very, very diluted. And McCarty was able to show that it could be destroyed not by enzymes that chewed up proteins, but by enzymes that chewed up DNA. Light-bulb moment. They couldn't see the DNA or even measure it properly, but they used independent lines of evidence that all said the substance was DNA. These included a range of chemical properties, and the fact that it didn't trigger an immune response when injected into animals – unlike proteins. This is excellent science. Independent lines of evidence leading to one overall conclusion are usually more reliable. It's unlikely that separate lines of evidence would all be wrong together.

They published their paper in 1944 and reported that the 'transforming principle' – meaning the chemical transforming a rough bacterium into a smooth one – was 'desoxyribonucleate'. They got the name slightly wrong, because it was actually known as deoxyribonucleic acid (DNA). But hardly anything had been written about it, so their mistake was understandable. I can hear Avery going, 'What's it called again? Must be desoxy, right?'

Avery, MacLeod and McCarty were modest in their conclusions, stating that 'the nucleic acids must be regarded as possessing biological specificity, the chemical basis of which is as yet undetermined.' But outside the confines of a scientific article, Avery was more bullish. A visiting scientist called Macfarlane Burnet, who himself would go on to win a Nobel Prize for discoveries in immunology, met with Avery. Later that evening, on a date with his fiancée, he told her that Avery had 'just made an extremely exciting discovery which, put rather crudely, is nothing less than the isolation of a pure gene in the form of desoxyribonucleic acid'. Now there is a rare example of scintillating conversation on a date.

The importance of the discovery was recognized immediately, with Avery winning the highest medal from the Royal Society in London: the Copley medal. Biochemist John Howard Mueller wrote: 'a polymer of nucleic acid may be incorporated into a living cell and will endow that cell with a property never previously possessed. When thus induced, the function is permanent, and the nucleic acid itself also reproduced in cell division. The importance of these observations can scarcely be overestimated.'

William Astbury, the first scientist to 'see' DNA, said in 1944: 'I wish I had a thousand hands and labs with which to get down to the problem of proteins and nucleic acids. Jointly those hold the secret of life, and quite apart from the war, we are living in a heroic age, if only more people could see it.' The general public were oblivious to all this excitement. There was, after all, a war going on.

Once the war ended, a frenzy of activity began. Some refused to believe that genes were made of DNA, still thinking that proteins were more likely. But to many, the sun had come out to illuminate the way. And it shone even brighter when Alfred Hershey and Martha Chase showed that when a virus infects bacteria, it is the viral DNA – and not the protein – that is injected into the bacteria, changing it dramatically. No protein goes in.

Some, however, weren't especially interested, and didn't seem to care what the chemical basis for inheritance might be. Delbrück (he of the 'green paper') said: 'And even when people began to believe it might be DNA, that wasn't really so fundamentally a new story, because it just meant that genetic specificity was carried by some goddamn other macromolecule, instead of proteins.' Goddamn indeed. And Hershey himself said: 'As long as you're thinking about inheritance, who gives a damn what the substance is – it's irrelevant.' Surely a curious thing to say, since he had identified the genetic material in his experiments as DNA.

This is not unusual in science, where people often stick to their own specialities and feel threatened by the new areas that come up. Geneticists weren't especially interested in the chemical nature of the gene, just in the science of heredity. Thankfully, many did give a damn.

It Has Not Escaped Our Notice . . .

The next piece in the puzzle came in 1948, when Colette and Roger Vendrely and André Boivin showed that a sperm and an egg each have only half the amount of DNA found in the fertilized egg. The 'two strands' idea was confirmed. And just like in the days of Miescher, biochemistry played a pivotal role.

Erwin Chargaff showed that whatever the source of DNA might be, the bases in DNA – described by Kossel and Miescher all those years ago, and given the letters A, T, C and G – had an interesting property. Using different sources of DNA, Chargaff found that the amount of A always equalled the amount of T, and the amount of C always equalled the amount of G. There was something funny there, he thought. But he knew it was probably important.

In 1952, Chargaff gave a lecture on this at the University of Cambridge, and there were two scientists in the audience, James Watson and Francis Crick. Watson had gone to Cambridge to work on DNA, and he and Crick, who had been a physicist, joined forces to try to work out the detailed structure of DNA. Crick had learnt some biology with Honor Fell in Strangeways Research Biology. The first time Crick and Watson met they discussed Schrödinger's 'What is Life?' lectures, which had had a huge influence on them – especially Crick. It was a bit like how new friends often bond over their record collections. They figured that if they could only find out what DNA looked like, that might reveal how DNA worked as the molecule of heredity, and put to bed once and for all the debate over whether the genetic material was a protein or a nucleic acid.

Watson and Crick were paid by the Medical Research Council in the UK, and another group similarly funded in

London was also working on the structure of DNA. In that group, a student called Raymond Gosling had managed to take a sharp picture of DNA using – guess what? – DNA from salmon sperm, just like Miescher had used all those years ago. Gosling's supervisor was Rosalind Franklin, who had refined the X-ray instrument used, and also had ideas about how to prepare the DNA sample – giving rise to the sharpest image of DNA seen up to that point. Franklin realized that the DNA in the resulting picture had the shape of a helix. Irish crystallographer J. D. Bernal, who was a huge influence on Franklin, called it 'amongst the most beautiful X-ray photographs of any substance ever taken'.

Franklin had shown huge intellectual ability as a child. Her aunt Mamie said of her when she was six: 'Rosalind is alarmingly clever – she spends her time doing arithmetic for pleasure.' She excelled in most subjects at school, with the exception of music. Her music teacher wrote to her mother at one point asking if Franklin suffered from hearing problems. Franklin carried out a PhD in physical chemistry at Cambridge, and learned to be a highly accomplished X-ray crystallographer in Paris.

Maurice Wilkins, who was born in New Zealand but whose parents were from Ireland, was part of the London research group, and Franklin gave him all her reports on the DNA work she had done when she moved labs, telling him to use it all as he saw fit. Watson came to visit one day, and Wilkins showed him the picture Gosling had taken (called *Photo 51*). That photograph, as well as other information from Franklin, led to Watson and Crick figuring out that DNA was a double helix.

It all suddenly became very clear. There are two strands in the DNA molecule (which was hinted at by the work on egg and sperm cells). The two strands are made of polymers of the A, T, C and G bases. If there is an A on one strand, it pairs with a T on the other. Think of them like Lego blocks clicking into each other. And if there is a C on one strand, that pairs with a G on the other. This is the reason why Chargaff saw that the amount of A always equalled the amount of T, and the amount of C always equalled the amount of G. The two strands click together like a ladder and then twist into a helix, which provides stability – just like a spiral staircase.

On 28 February 1953, Crick and Watson went for a lunchtime pint in the Eagle pub in Cambridge and announced that they had 'discovered the secret of life'. I wonder what the locals thought.

They had a mechanism for how DNA copies itself, passing on the information to the next generation. When a cell is dividing, the two strands separate, and then a new strand is made to line up with each of the separate strands, with the Lego-block bases clicking into place in the new pairs – and hey presto, the DNA has been copied. One copy goes into one cell, and the other copy goes into the other cell. And it's all driven by the double-stranded DNA and the pairing of bases. If the DNA is being copied when a cell divides, it must contain the recipe to make a new cell. But neither Watson and Crick nor anyone else at that stage knew what that recipe was.

In 1953, the work of Watson, Crick, Franklin, Gosling and Wilkins was published in three back-to-back articles in the world's leading science journal, *Nature*. Franklin had already submitted other work to the journal *Acta Crystallographica* describing the double-helical backbone of DNA. The Watson and Crick article contained the sentence: 'It has not escaped our notice that the specific pairing we have postulated immediately suggests a possible copying mechanism for the genetic material.' Oswald Avery is thought not to have seen these papers. He had retired in 1948 and gone to live with his brother in Nashville, and died two years later.

Crick wrote to Schrödinger in Dublin, enclosing copies of the *Nature* articles: 'Dear Professor Schrödinger. Watson and I were once discussing how we came to enter the field of molecular biology and we discovered that we had both been influenced by your little book, *What is Life?* We thought you might be interested in the enclosed reprints – you will see that it looks as though your term "aperiodic crystal" is going to be a very apt one.'

Nobel Prize Stampede

Next was a lot of hard work linking DNA to protein production, with a lot of people involved. Crick himself had come up with the idea of what would become the central dogma of molecular biology: DNA makes RNA, and RNA makes protein. It was known that genes in DNA must lead to proteins being made, since it is the proteins that do all the heavy lifting of life and DNA somehow must instruct their production. And scientists had begun to realize that RNA, a nucleic acid like DNA, was an intermediary of some kind.

And then three scientists, Marshall Nirenberg, Har Khorana and Robert Holley, broke what had become known as the genetic code. If we go back to the four bases in DNA, they are not only the key to the linking of the two strands of DNA; they tell the cell how to make proteins. They occur in sets of three, and each set instructs the cell to put an amino acid in place. Three of the bases govern which amino acid will be in place in the protein. It's all done via the intermediary of RNA, in a structure called the ribosome. In the ribosome, the amino acids line up and then link together into a protein, which then folds into shape.

It took a huge amount of biochemistry, cell biology and molecular biology to figure all that out. And it was the work of Nirenberg and his colleagues that revealed it. Nirenberg had developed an interest in biology as a child. He obtained a master's in zoology from the University of Florida, working on the taxonomy of caddis flies. He switched to biochemistry in 1959, and began working on DNA, RNA and protein, trying to link them all. He made an RNA molecule with only one base, U, repeated ad infinitum: UUUUUUUU . . . He added it to an extract from the bacterium *E. coli*, then added one radioactively labelled amino acid (which could be traced, because it's radioactive) and 19 unlabelled amino acids. He tried this with all 20 amino acids, using one radiolabelled amino acid each time. He could track each radioactive amino acid in turn, and what he found was that only one radiolabelled amino acid of the 20 formed into a protein. That was an amino acid called phenylalanine.

Nirenberg had made the first protein (or more accurately peptide, because it wasn't a complete protein) from an RNA molecule. The *E. coli* extract contained ribosomes which did the work. It meant that AAA in DNA coded for phenylalanine, acting via RNA – which had U instead of T (which remember, occurs in DNA). The AAA in DNA made the UUU in RNA, again by base pairing. And in this elegant experiment, Nirenberg and his colleagues cracked the genetic code.

Nirenberg first made a presentation on this in 1961 at a

conference in Moscow, to a small audience. At the end of his talk, another important scientist working on DNA, Matthew Meselson – who with Franklin Stahl had worked out the chemical basis for how DNA copies itself (no mean feat in its own right) – went up to him and spontaneously hugged him. Francis Crick was at the conference and Meselson told him about Nirenberg's discovery. Crick insisted that Nirenberg speak again, but this time to a much larger audience.

This time, more than a thousand people turned up to listen, and they were electrified by what they heard. It was as big as the Sex Pistols gig at the Manchester Lesser Free Trade Hall that led to lots in the audience forming punk rock bands of their own. All the key findings were now in place.

The process of copying the information from DNA into RNA is called transcription, because the DNA and RNA languages are very similar. The turning of the information in the RNA into protein, however, is called translation, because the language of proteins – being made of amino acids – is very different from the language of DNA and RNA, which are made of nucleotides

To put it simply, DNA, the molecule first seen in pus by Miescher all those years ago, is what genes are made of. It can copy itself when a cell divides, because it has two strands wound up into a double helix. The strands separate and make copies of themselves. And then, when DNA wants to express itself ('gene expression' is the term that's used), it encodes for RNA, which then encodes for proteins. That's it. The secret of life solved. And all living things on Earth do exactly that. Even viruses, responsible for many diseases, contain just a few of their own genes and proteins and rely on their host cells to provide the machinery and energy to make more virus. Salmon, peas, sea urchins, bacteria . . . you. They are all descended from the first cell that figured all this out, billions of years ago. Complex chemistry – but chemistry just the same.

Lots of Nobel Prizes were awarded for all of this. Watson, Crick and Wilkins shared one in 1962. Sadly, Rosalind Franklin had died, otherwise she would have won one too (the rules preclude giving a Nobel Prize posthumously) – most likely in Chemistry, as Watson suggested, as only three can win a single Nobel (yet more rules . . . scientists love rules), and the others had won it for Medicine or Physiology. That Nobel Prize really upset Chargaff, the scientist who had shown that the amount of the base A always equals T, and the amount of C always equals G, and he shut down his lab and wrote to many scientists, complaining bitterly. Ah! Human nature yet again. Nirenberg and colleagues won theirs in 1968. But there was yet more to come.

Cut All the Whacko Stuff Out of It

Chemist Fred Sanger figured out how to read the order of the bases in DNA, and also the order of the amino acids in proteins, which confirmed everything. Sanger had obtained a PhD at Cambridge, working on the amino acid lysine in nutrition. He continued working on amino acids, and in 1952 figured out the order of them in the hormone insulin. Proteins were known to be polymers, made up of strings of amino acids that fold into shape. This was a big advance, as until then proteins had been thought of as amorphous. In the 1960s Sanger shifted his attention to RNA and then DNA, working out a chemical method to get the sequence of the bases. He won two Nobel Prizes – one for each of these discoveries – then retired at the age of 65 and took up gardening with gusto. He turned down a knighthood, saying 'a knighthood makes you different, doesn't it, and I don't want to be different.' He described himself as 'just a chap who messed about in a lab.' Some messing.

The discoveries about DNA kept coming. Again, new techniques were important. Hamilton Smith and Thomas Kelly discovered an enzyme that could chop up DNA in very specific ways. Paul Berg figured out a way to splice different pieces of DNA together. All this became important for a technology known as recombinant DNA. You could splice the gene for, say, growth hormone or insulin into bacteria, and get the bacteria to make the hormone – which could be used to treat patients with dwarfism or diabetes, ushering in the era of DNA technology and genetic engineering.

In the 1980s, Kary Mullis figured out a way to hugely amplify RNA molecules, which made studying them much easier. Mullis obtained a PhD in biochemistry in 1973 for work on a bacterial protein involved in iron transport. He almost failed, and it's been said that his friends helped him rewrite his dissertation, convincing him to 'cut all the whacko stuff out of it'. He left science after that and ran a bakery for two years. He then went to work for a biotechnology company called Cetus, and in 1983 discovered a technique called the polymerase chain reaction (PCR). He said that he made the breakthrough on 16 December, and his then girlfriend broke up with him on the same day. He recalled that the discovery didn't make up for the pain he felt. Mullis told people that he had taken a lot of LSD as a student at Berkeley and that it helped him come up with the idea of PCR. He also worked for the defence team during the O. J. Simpson trial. The prosecution painted him as a drug-addled surfer dude who had denied that HIV causes AIDS. Who says scientists are boring nerds?

The PCR technique revolutionized molecular biology, as it was possible to amplify huge amounts of RNA into DNA to study and manipulate. This is the gold standard test for detecting the RNA in COVID-19, and has many other uses. Mullis himself founded a business to sell pieces of jewellery containing the amplified DNA of deceased famous people such as Elvis Presley and Marilyn Monroe. You can't fault him for his entrepreneurial skills. He was however resentful of his employer Cetus. They paid him a $10,000 bonus, but sold the patent for PCR to the pharmaceutical company Roche for $300 million.

All these discoveries ushered in the era of what is called genomics, which is the study of all the DNA content in a cell. And this led to the entire human genome being sequenced in 2000. We now had the full recipe to make a human, and we could predict what all the proteins would be from the DNA sequence based on the genetic code – because remember, that code predicts the order of amino acids in a protein. This could be done for any organism. Computers became especially important to analyse all the data. The human genome has 3.2 billion bases all in sequence over the chromosomes. Try figuring that out

Genomics allows us to spot mutant genes that give rise to disease, and this might lead eventually to cures. By 2018, 100,000 human genomes had been sequenced. A single mutation in one base out of the 3.2 billion can cause devastating genetic diseases. This is because that base change can lead to an amino acid change in the protein being encoded, which can inactivate it or make it overactive. There are thousands of examples of this happening. For example, in cystic fibrosis, a genetic change alters a protein involved in controlling the amount of salt in the lungs. When this is broken, the lungs can't handle salt as well, and lung damage results. Cancer is usually caused by a chemical (called a carcinogen) altering a base in DNA which in turn leads to change in an important protein, either activating it to cause tumours to grow (for example, the protein Ras) or inactivating it if the protein is a tumour suppressor (for example, the protein p53). Identifying and analysing these mutations in patients is a massive task. It is thanks to the development of powerful computers that we can now make use of genomics to spot mutant genes.

By 2009, Ada Yonath, Venki Ramakrishnan and Thomas Steitz had solved the structure of the ribosome, the big machine that reads off the RNA sequence and makes all the proteins in your body. This was a tour de force of structural biology that goes back to the days of Rosalind Franklin. When I invited Ada to come to Trinity College Dublin and give a talk, she told the story of how she did it, leading to the Nobel Prize. She said it gave her a thrill, but not as big a thrill as being with her grandchildren – who are after all made of billions of ribosome factories (she didn't say that, but I was thinking it).

And then, as we read in the chemistry chapter, Jennifer Doudna and Emmanuelle Charpentier discovered CRISPR, which is a way to edit genes accurately. A specific sequence is targeted with a complementary sequence – leading to the DNA being cut or altered at that precise spot. I met Jennifer in 2016 when we were both bestowed with the honour of Fellow of the Royal Society. I asked her what she thought when she discovered CRISPR. She said she knew it must be correct because it relies on the pairing of bases in DNA. That pairing was ringing through all the way from Chargaff.

CRISPR might be a way to fix the broken genes that cause diseases. In 2020, it was used for the first time to treat a blinding disease called Leber's congenital amaurosis. It holds great promise in terms of treating other genetic diseases where DNA is mutated – and there are loads of them.

Out of Africa

Because it is a very chemically stable molecule, we can even study the DNA from people and animals that died a million years ago. We can effectively look back in time. DNA has been examined from the bones of Neanderthals, an extinct subspecies of humans who lived in Europe and parts of Asia until about 40,000 years ago. Our own subspecies of human, *Homo sapiens*, encountered them when we began migrating out of Africa between 60,000 and 90,000 years ago. It wasn't known whether we bred with Neanderthals, but DNA analysis has revealed that we did. By comparing the DNA sequences from Neanderthals to our own, we can see that we inherited gene variants for skin colour, bone density, lung capacity and parts of our immune system from Neanderthals. That could never have been worked out from archaeology alone; it took DNA analysis to learn about our ancestry.

We've come an awful long way since Miescher first described DNA in pus taken from bandages. Even bandages themselves have become really high-tech. They can kill bacteria, help wounds to heal, and even detect antibiotic-resistant bacteria. But we're not talking about advances in bandage technology here; we're talking about DNA. We now know how DNA works. We've revealed the stuff of life itself, something we had no real notion of even 60 years ago. We know how evolution works – through random DNA mutation. This is because the copying of DNA sometimes makes mistakes. Without those mistakes there would be no us, because evolution wouldn't happen. Through DNA, we can explain what Darwin discovered. We have figured out how we're all descended from a cell that arose probably 4 billion years ago or more, which had DNA as the recipe for life. The same DNA that is in us now. We know how DNA can go wrong in disease. And hopefully we can come up with ways to fix it through what's called gene therapy. We have the power to alter DNA, which was previously only achievable through selective breeding or evolution. We can even create synthetic life – making DNA in a test tube and booting up life by inserting it into a bacterium. We have devised the PCR test, and can use RNA or DNA to make vaccines. The prospect of the prevention or treatment of so many diseases is within our grasp – and it's all because of pus-filled bandages.

7. Medical Love Song

When Charles Darwin dropped out of medical school in Edinburgh, medicine's loss was biology's gain. Several of the people who made discoveries in biology and DNA were medical doctors or medical school dropouts. Perhaps Darwin and many others, having been exposed to subjects like botany or chemistry, couldn't resist the pull of science. I have collaborated with many medical doctors in my career and most of them were drawn to research because they wanted better ways to treat their patients.

My mentor in Cambridge, Jerry Saklatvala, was a rheumatologist, and told me that he went back to the lab because he got tired of recommending to women who had arthritis in their ankles to change their shoe size. He felt that surely there should be better options, and he was one of the key scientists who uncovered the role of the inflammatory protein TNF in rheumatoid arthritis, the blocking of which has brought benefits to millions of patients.

At the time of Darwin, medicine had scarcely made any impact on human health in the past 2,000 years. So, there was a pressing need to make scientific discoveries that might help develop treatments for the diseases that had afflicted humanity for many thousands of years. The truth of it was that most physicians generally did more harm than good when treating their patients before modern science-based medicine emerged in the twentieth century. The main treatments they recommended, in Western medicine at least, were a lot of bleeding and vomiting. Too much bleeding is obviously very bad for you, and too much vomiting can cause dehydration as well as injury to the oesophagus. Although this could be the result of a 'fun' night out, it was hardly an inducement to visit the doctor. And medicine wasn't scientifically based, meaning coming up with a hypothesis, testing the hypothesis with as much rigour as possible (especially involving multiple lines of evidence as we saw in previous chapters) and then drawing a conclusion that was reproducible (i.e. true). Sadly, much of that was lacking in medicine until the nineteenth century.

Modern medicine as we now experience it was founded on the biological knowledge that had built up as humans wondered about the nature of disease. Many of the advances you read about in the chapters on chemistry, biology and DNA were important precursors for what happened in medical research from the nineteenth century on. The results of the application of the scientific method to understanding and treating diseases have been a spectacular success, although it was a long time coming.

Plague Teeth

Strange as it may seem, diseases and civilization go hand in hand, but it may not always have been that way. Ten thousand years ago humans were hunter-gatherers. We lived and travelled in small groups, and foraged for food, probably having quite a varied diet. Disease would have been a result of injury, malnutrition if food was scarce, exposure to parasites in drinking water, and infections in scrapes and wounds or caught from wild animals and intermittent contact with other tribes. Disease may well have been a relatively rare event.

But when agriculture started and we began to live in fixed settlements, this gave rise to all kinds of problems. Domesticated animals became sources of infection, and the higher population density facilitated spread through a community. Many infectious diseases that have come to literally plague us came from the animals we had close contact with. The Black Death hopped into people from the fleas on the backs of black rats that were infected with the bacterium *Yersinia pestis*, named after the Swiss doctor Alexandre Yersin who first described it. Archaeological finds of teeth from Bronze Age humans show DNA evidence of Yersinia infection. Plague has been with us for millennia.

Many flu pandemics started in pigs, and birds such as poultry. And of course, the most recent pandemic, where the SARS-CoV-2 virus most likely jumped from a bat into a human, possibly via an intermediary species – the most recent contender being the racoon dog, of all things. The technical term for diseases that jump from other species is 'zoonoses' – sounds almost pleasant, although a zoo containing fleas, rats, pigs, chickens and racoon dogs might not be such a nice day out. Especially with all the diseases.

Along with the risk of infectious diseases posed by close contact with animals, paradoxically people were also probably malnourished once we adopted agriculture. Although crops increased the quantity of food, it became less diverse because people lived off one or two crops. Malnutrition weakens the immune system and so makes infection worse. Settlements also meant waste piled up, which attracted insects that spread more diseases – there is nothing flies and mosquitoes like better than a big, warm, moist pile of human and animal excrement. And so disease became a feature of 'civilized' life, and this affected our chances of living a long and healthy life.

Thousands of years ago, if you managed to make it past childhood, you might expect to reach your fifties, compared to around 80 years in Europe and the US if you're alive today. This ~60 per cent increase in life expectancy is because of a revolution in our chances of surviving childhood and some spectacular advances in the treatment of specific diseases, and we are probably now entering another advance which will prolong the average healthy lifespan even further. More on that later.

There's no accurate data on how long people lived before agriculture was widely adopted – it was a bloody long time ago – but anthropologists have examined today's hunter-gatherers, such as the Aché people of Paraguay and the Hadza people of Tanzania. Babies in those groups have a lower chance of survival – a Hadza boy has only a 55 per cent chance of making it to the age of 15, while for an Aché boy it is 71 per cent. In comparison, the CDC reports that a child in the US today has a greater than 99 per cent chance of making it to 15.

But it's something of a myth that, before the medical advances of the twentieth century, people routinely lived lives that were 'brutish and short'. One study spanning 1200–1745 has shown that if you made it to 21, you could expect to live to between 62 and 70. However, being pregnant in that period was definitely riskier, mainly because of infectious diseases. Women undergo a series of changes to their immune system during pregnancy that are necessary to assist in implantation of the fertilized egg, development of the foetus, and to prepare for delivery. Filthy doctors and midwives were often delivering a large dose of infection as well as children. Childbirth itself was dangerous, as was early childhood because of an under-developed immune system.

Even now, a child is two- to three-fold more likely to die of influenza than an adult is, and a study in England and Wales has shown that one in five childhood deaths are caused by infections. But making it past childhood in the distant past meant you would go on to have a reasonable lifespan.

By the late Victorian period, which we associate with the gloom, pollution and general misery of the novels of Charles Dickens, records show that if Oliver Twist or Tiny Tim made it to adulthood (avoiding Fagin and Ebenezer Scrooge), life expectancy wasn't that different to today. A five-year-old girl could expect to live until 73, while a boy might reach the age of 75. Why the difference? Well, it turns out that girls were inclined to be fed less than boys, because it was assumed they didn't need as much nutrition. This has been shown to weaken bone development, including the pelvic bone, which created further risks during childbirth.

The reasons then for the vastly improved life expectancy during the twentieth century include better nutrition and hygiene, but also antibiotics and vaccines (more on that later). But health isn't just about how long you live – your lifespan. It's also about health span – how long you can live a *healthy* life for. And that involved the diagnosis and treatment for diseases that might afflict you.

The Squealing Pig Trick

The earliest medical prescriptions we know of are from Mesopotamia around 3000 BCE, and involve 15 remedies which include snakeskin, turtle shell, myrtle, thyme and willow, although curiously they don't document the diseases that would have been treated. There are various Egyptian medical papyri that have descriptions of ailments with associated cures, as well as anatomical observations. The Ebers Papyrus, which we read about in chapter 5, dates from around 1550 BCE and covers treatments for many diseases, including digestive diseases, gynaecological conditions and even migraines. The Edwin Smith Papyrus – named after the archaeologist who purchased it in the 1860s – dates from 1600 BCE and so is the oldest known surgical text. It describes 48 cases of trauma, each with a physical examination, diagnosis, treatment and prognosis. The Kahun Gynaecological Papyrus exclusively deals with women's health, including fertility, pregnancy and contraception, and dates from around 1800 BCE. The earliest known physicians are Egyptian: Hesy-Ra, who was born around 2600 BCE and was known as 'Chief of Dentists and Physicians', and Peseshet, born in 2500 BCE, who is the first known woman physician, known as 'Lady Overseer of the Lady Physicians'.

Many ancient cultures had healers, who used herbs, animal products, enemas, poultices and ointments of various kinds. In early medicine you had to 'suck it and see'. This is not unique to humans: chimpanzees have also been known to eat plants that rid them of worms and apply insects to wounds to help healing. How they learned how to do this is still not understood, but younger chimps appear to acquire this behaviour from adults. How the insects heal wounds is not known, but it could be due to multiple antibiotic compounds that many insects have been found to contain, preventing infections in themselves.

The Greeks provide us with the earliest accounts of what would become modern medicine. Hippocrates compiled 70 medical works and gave us the Hippocratic Oath. In its original form, it was sworn to Apollo, who was god of both medicine and music, and indeed music was often used in the ancient Greek healing process. In the oath, after mentioning Apollo and other gods and goddesses, the physician then swore that he would hold his teacher 'equal to' his own parents, and that if the teacher needed money, the newly minted doctor would always share his. Clever, right? I should get my own students to swear such an oath to me.

A key part of the oath was of course to do no harm, and to maintain patient confidentiality. But it also covered avoiding the seduction of women and men, free or enslaved, when performing home visits, and bound its physicians to leave surgery to the surgeons – the first example of a formal medical specialization. Though, as with the doctors, at this time your surgeon might well have done more harm than good, since so little was known about surgery. The instruction to physicians to avoid surgery was probably also an acknowledgement of the limitations of surgery.

Hippocrates described how illness can be explained by an imbalance in the four 'humours' or fluids: blood, phlegm, black bile and yellow bile. There is no basis for such imbalances in diseases, but given the careful recording of symptoms that went along with it, we are on our way to proper scientific medicine. He also classified diseases as either acute, chronic, endemic or epidemic, and used such terms as 'exacerbation', 'relapse', 'resolution' and 'convalescence'. Greek physicians often prescribed changes in diet or lifestyle to combat diseases. They also practised bloodletting. As the blood flowed, this was thought to rebalance the humours – a practice which persisted up to the twentieth century, even though there was no evidence for it being useful and, if anything, it probably did harm. The only exceptions are for some rare blood diseases such as haemochromatosis, where phlebotomy (bloodletting) is used to reduce pathological levels of iron or red blood cells.

During the second century CE, another Greek (but who worked mainly in Rome), Galen, had a huge influence on medicine, and this lasted until the 1500s. Wouldn't it be great if some of my discoveries in the immune system are still being talked about 1,700 years from now? His findings often came from the dissections he carried out on different animals that were anatomically similar to humans. One dissection he performed in public that became famous was the 'squealing pig'. He would cut open a pig and, while it was squealing, he would tie off the pig's vocal cords, showing that they were responsible for the noise. People paid to observe this particularly perverse party trick, but I guess there wasn't much entertainment back in those days.

Many of his earlier anatomical reports had been based on the dissection of Barbary apes, but he switched to pigs because the facial expressions of the apes disturbed him. He became the physician to the gladiators of the High Priest of Asia. He was chosen because he had eviscerated an ape and asked the other physicians to repair it. They couldn't, and he successfully performed the surgery himself. Performing dissections on dead human bodies was strictly banned in ancient Rome but Galen encouraged his students to examine dead gladiators or bodies washed up after shipwrecks.

He also had first-hand experience of the Antonine Plague, which struck Rome in the year 165 and again in the winter of 168–69. It's also known as the Plague of Galen, because of the careful notes he kept on the victims. The plague is thought to have killed up to 5 million people, and some have estimated that it caused the deaths of half of the entire population of the Roman empire. And we think we had problems with

COVID-19? From Galen's description, the Antonine Plague was caused by smallpox, a disease that would ravage Europe until the discovery of the protective effect of vaccination by Edward Jenner some 1,650 years later – more on that later. The awfulness of smallpox, with 50 per cent mortality in the Antonine Plague, and the power of vaccination – the smallpox vaccination has now eliminated smallpox around the world – show us just how important science is.

Von Gebweiler's Skeleton

Galen had a significant influence on Islamic physicians, including the Persian polymath Avicenna, who wrote a major medical text called *The Canon of Medicine* that dates from 1025. Avicenna had memorized the entire Quran by the age of 10, and learned Indian mathematics from an Indian greengrocer whose name is still known a thousand years later – Mahmoud Massahi.

Avicenna's book is an overview of medical knowledge from multiple sources, including Galen, and also works from Persia, China and India, so it's something of a tour de force. It was a standard medical textbook up to the eighteenth century in Europe. It opens with essays on anatomy and the body. After a diversion to outline the elements of cosmology, it then provides a diagnosis and treatment of diseases (drawing heavily on Galen's four humours) and provides a list of treatments for disease in alphabetical order. Several specific neurological disorders are covered, including epilepsy, stroke and vertigo, but also 'love sickness' and 'melancholia'.

Avicenna wrote more than 450 works, of which 240 have survived. He suffered numerous illnesses himself, which were attributed to his work rate, but he refused to slow down, saying, 'I prefer a short life with width, to a narrow one with length.'

Galen's impact can also be seen in the works of Vesalius, a sixteenth-century physician from what is now Brussels, who described in detail the anatomy of the brain and most organs. His actual name was Andries van Wezel, but it was common for smart people at that time to Latinize their names – Linnaeus, who we met in the zoology chapter, was actually Carl von Linné. How Andries got from Wezel to Vesalius isn't known, but perhaps he changed his name because it sounded too much like 'weasel'.

Vesalius was so smart that when he graduated in medicine from the University of Padua he was immediately offered the chair of surgery and anatomy. He spent a lot of his spare time examining bones in cemetery vaults. Whatever turns you on.

In 1543, he famously performed a public dissection of an executed prisoner, Jakob Karrer von Gebweiler, to much shock and awe. This drew a large audience, as von Gebweiler was a notorious criminal who had stabbed his wife after she discovered he was a bigamist. Afterwards Vesalius assembled von Gebweiler's skeleton, and it is still on display at the University of Basel and is the world's oldest surviving anatomical preparation. In the same year, he published the anatomical book which he is most famous for: *De Humani Corporis Fabrica.* This was in the same year that Copernicus published his great book *De Revolutionibus Orbium Coelestium*. Some year. You needed to know your Latin in those days, otherwise you couldn't have read these books – and you wouldn't have been able to give yourself a cool Latin name.

In English, the title of Vesalius's book is *On the Fabric of the Human Body*, and it caused a sensation. This book is seen as the start of medicine becoming more scientific. He specifically stated that it was an advance of the work of Galen – the first major advance in 1,400 years. In the book he corrects over 200 mistakes made by Galen, which just shows you: anyone can make a mistake in a book – even me. I've tried very hard to fact-check this book. Ten points to the person who finds the most mistakes. One thing Vesalius corrected concerned blood circulation. Galen had stated that the right and left ventricles of the heart were connected by tiny openings to allow blood to circulate. Vesalius tried hard to find the openings but couldn't. This flew in the face of 1,400 years of other anatomists stating that they could observe them, a perfect illustration of 'group think'. Galen had also stated that the major blood vessels originated in the liver, which Vesalius also questioned. Galen's mistakes most likely came from the fact that he mainly drew conclusions from dissected dogs and monkeys. It's a lesson we must remember today, where often findings are made in model organisms like mice, which don't necessarily translate to humans.

The book wouldn't have been possible were it not for the artistic advances that had happened in the Renaissance, and the printing process allowing for woodcut engravings – these advances gave us highly detailed illustrations that could be mass-produced. This illustrates how you often need a few things to go right in science for there to be an impact – not just the discovery itself. Vesalius gave the first copy to his patron, the Holy Roman Emperor Charles V, but had it bound in purple velvet, with many hand-painted illustrations. My publishers are yet to agree to a special edition of this book with an emerald-green velvet cover and gold inlay.

More than 700 copies of the great book survive, and one of them is held at Brown University in the US and is bound in tanned human skin. Clearly someone had skin in the game, but whose skin was used is not known. The book was donated by Albert Lownes as part of his collection of books of significance, which he defined as 'books that have changed the world or man's way of seeing it. Significance also meant books that I found interesting.' The collection contains over three-quarters of those texts recognized by scholars as the 'great books' of science published since the middle of the fifteenth century – libraries are cool places.

Vesalius came under constant attack, both from professional rivals who were incensed that he had challenged the great Galen but also the clergy because dissecting human bodies was seen as unethical. One rival even stated that the human body had changed since the time of Galen, and so Vesalius was wrong in his corrections to Galen's work. Sometimes it can be hard being a scientist. How do you counter that kind of thing?

He went on a pilgrimage to the Holy Land, possibly as penance for dissecting a living human body, although it could also have been a way to escape his post at the Spanish court. Though going on a pilgrimage to escape your job seems rather desperate. On the pilgrimage he was shipwrecked on the Greek island of Zakynthos and died there at the age of 49. He was so poor that a benefactor had to pay for his funeral.

The Witch's Toad

The scientific basis for medicine really took off when English physician William Harvey described the circulation of blood. In one fell swoop he described the function of the heart and lungs. Harvey was born in Folkestone in Kent and studied medicine at the universities of Padua and Cambridge, graduating in 1602 at the age of 24. In Padua, he mightily impressed his examiners, with one biographer recording that he 'showed such skill, memory and learning that he had far surpassed even the great hopes which his examiners had formed of him'. He spent most of his career working in St Bartholomew's Hospital in London. In 1615 he was appointed 'Lumleian lecturer', a post founded by one Lord Lumley with the purpose of 'spreading light' about anatomy in England. Harvey's notes survive, and he clearly laid out the purpose of the lectures: 'To show as much as may be at a glance, the whole belly for instance, and afterwards to subdivide the parts according to their positions and relations.' He published his book on the circulation of blood in 1628. Yet another text in Latin: *De Motu Cordis* (*On the Motion of the Heart*). His ideas on circulation were again heavily criticized by other doctors.

Harvey was known for criticizing allegations of witchcraft, and for his trouble he got a job examining women who were accused of being witches. This wasn't uncommon and occurred for various reasons, mainly to do with the oppression of women. The Catholic Church taught that women were the weaker sex, having already been tempted by the devil in the Garden of Eden to eat the apple even though God had forbidden it. If a woman stepped outside her traditional role of mother and wife, or if she was even simply outspoken, she risked the accusation of being a witch. Between 1612 and 1634, many instances of witchcraft were reported in the Pendle area of Lancashire. Harvey was involved in a case of 17 women who were found guilty of witchcraft, and all on the testimony of an 11-year-old boy who later confessed to making up his story to make money.

Harvey examined four of the women for physical signs of witchcraft such as the mark of the devil, or an extra teat capable of suckling a witch's familiar – supernatural entities believed to help witches in their practice of magic. Familiars were thought to be able to appear in the form of an animal, and were most commonly small animals such as cats, dogs, frogs or toads. Harvey didn't find any evidence and the women were pardoned by King Charles I.

He was also sent one day to examine a woman in Newmarket who was accused of being a witch. To win her confidence he said he was a wizard and asked if she had a familiar. The woman put down a saucer of milk and a toad came to drink it. Now that must have been cool. Harvey then asked the woman for a glass of ale, and she went to fetch it. He examined the toad closely, killed it and dissected it. He concluded that the toad wasn't supernatural in any way. The woman was very upset that he had killed her pet toad. But he calmed her down by telling her that he was the King's Physician sent to discover if she was a witch.

Along with anatomy, progress was also happening in the effort to discover medicines to treat diseases. Throughout human history, and in every culture, there have been herbal remedies for all kinds of diseases, some of which are still in use today. In China it was ginseng, which was used to increase strength and treat erectile dysfunction, and cinnamon for back pain, asthma and arthritis. Indians loved nutmeg for nausea, stomach cramps and diarrhoea. Egyptians highly rated pomegranate for tapeworms, aloe for burns, and garlic for general malaise. The Celts had many herbal remedies, including juniper berries for epilepsy and sage, which was considered a cure-all. Sick people were advised to keep away from 'dogs, fools and talkative noisy people'. Excellent advice to this very day.

Chemists Rarely Move

The nineteenth century marks the rise of modern medicine. Active ingredients were prepared from many plants, including quinine from cinchona bark (for malaria) and anti-inflammatory salicylates from willow bark (aspirin is a type of salicylate); and so, pharmacology as a science began.

The first pharmacies were in the Middle East, and there are records of their arrival in Baghdad in 754. By the ninth century these were regulated by the state, illustrating how even then there were concerns about quack remedies being sold that either didn't work or were harmful. In Europe, pharmacies began to appear in the twelfth century. The Town Hall Pharmacy in Tallinn in Estonia dates to 1422 and is the oldest continuously run pharmacy in the world – still operating in the same premises. Maybe James Joyce got it right when he wrote in *Ulysses*: 'Chemists rarely move. Their green and gold beaconjars too heavy to stir.'

Over the course of the nineteenth century, developments in pharmaceuticals came thick and fast – sometimes literally, since pharmacists loved making electuaries (look it up, it's also in *Ulysses*), tinctures and syrups. Advances in chemical methods allowed for more effective purification of active ingredients. Morphine was first prepared from a resin from the poppy (hence the term 'opioid', from opium) by the German pharmacist Friedrich Sertürner in the early 1800s, and is considered the first active ingredient prepared from a plant. It was used for pain and to help with sleep. He named it after Morpheus, the Greek god of dreams.

Sertürner published a paper on morphine's isolation, crystallization, crystal structure and properties, which set the standard for how to report on the preparation of drugs. He tested the preparation first on stray dogs and then on himself to treat a terrible toothache. He then performed a simple clinical trial on himself and three other people to establish the optimal dose, in which one-fourth grain (30 milligrams) of the drug induced a happy, light-headed sensation, the second dose caused drowsiness and excessive fatigue, while the third caused participants to become confused and somnolent. We now know that morphine works by stimulating what are called opioid receptors in the brain, blocking pain signals. Sertürner suggested that a 15 milligram dose was optimal – a dose that is still used today. He was granted the title 'Benefactor of Humanity' for his work, clearly justified given the benefits of morphine in pain relief, but he also became addicted to his own discovery. It took another 200 years before Tony Montana in the movie *Scarface* was told 'don't get high on your own supply' – advice he failed to follow and that came far too late for Sertürner. Sadly, we have Sertürner to thank for the current opioid crisis in the US, since oxycodone, which is a major factor, is an opioid.

In the 1890s physiologists began crushing up various organs, administering them to patients, and seeing what activity they could find. Extracts from the adrenal gland had all kinds of effects on the body, including the raising of blood pressure via the constriction of blood vessels. Surgeons used the extracts to maintain blood pressure during surgery, or if a patient went into shock, where blood pressure can fall dangerously. This led to the purification in 1897 of an active substance that was named epinephrine. This was developed by the pharmaceutical company Parke-Davis, who trademarked it as Adrenalin, and it was initially used to treat asthma attacks because it could relieve the constriction of the airways. It's still used for severe allergic attacks in the form of an EpiPen. It was also useful as a nasal decongestant, though this use was limited because it had to be injected, and so an oral form was sought, leading to the discovery of ephedrine in 1885 by Japanese chemist Nagai Nagayoshi. This was also used for asthma and led to the discovery of methamphetamine, a derivative of ephedrine.

In 1929, one Gordon Alles – after testing amphetamine on guinea pigs – became his own guinea pig and took a huge dose himself. It wasn't that unusual for medicinal chemists to sample their own wares at that time, dangerous as that might be. After noting heart palpitations and a rise in blood pressure he had 'feelings of well-being' and considered that he was unusually witty at a dinner party (I wonder if the guests felt the same), after which he had trouble sleeping. It was initially sold as a decongestant, used to fight fatigue in soldiers, and then marketed as an antidepressant up until the 1960s. The misuse of amphetamine as 'speed' and its relatives such as 'crystal meth' is still a problem, and medical uses of amphetamines are limited to specific examples such as attention deficit hyperactivity disorder (ADHD). Amphetamines work by stimulating the release of the neurotransmitter dopamine and regulating another neurotransmitter, glutamate. We'll read more about these in the chapter on the mind.

Doctors also found that extracts from the pancreas could have various effects on the body. In 1889 it was shown that diabetes could be triggered in dogs if their pancreas was removed. Diabetes is characterized by high blood sugar after a meal, which leads to all kinds of problems and if left untreated can be fatal. Canadian Frederick Banting and his student Charles Best used pancreatic extract to treat the dogs whose pancreas had been removed. Others had shown that a part of the pancreas called the islets of Langerhans (which to me always sounded like an exotic destination) was needed to make a protein to control blood sugar, and had named it insulin. Efforts to extract insulin had failed because enzymes in the pancreatic extract destroyed it. Banting and Best came up with a way around this problem and made insulin in bulk. Cows and pigs became the main source of insulin for diabetics up to the late twentieth century, when genetic engineering was used to make insulin in bacteria.

Banting won the Nobel Prize in Medicine or Physiology in 1923 for these discoveries. He was 32 at the time and is still the youngest recipient in Medicine/Physiology. He shared the prize money with his student Best. Prior to their discovery, people diagnosed with diabetes lived only a few months.

Banting's main hobby was painting. He must have been a very frugal man because many of his paintings were done on the cardboard inserts dry-cleaners put into his shirts. He particularly enjoyed painting arctic landscapes, and at the time of his death in a plane crash in 1941 – at the young age of 49 – he was one of Canada's best-known amateur painters.

There is still no cure for diabetes. In 1989, a 'Flame of Hope' was lit in Banting's home town of London, Ontario, and it will be extinguished by the researchers who discover the cure. Recent progress on using cell therapy suggests that the science that will put the flame out is already underway.

Wash Your Hands!

Meanwhile, many other diseases were rampant, and efforts to prevent or treat them continued apace. For example, in 1900, pneumonia, tuberculosis and diarrhoea were the leading causes of death in the US, and infant mortality from infection stood at 10 per cent. Improvements in public health and hygiene had a major impact, and although the idea that infectious diseases were spread by microscopic creatures that couldn't be seen gained traction (microscopes certainly helped here!), it took some time to find drugs to treat infectious diseases.

In the 1840s, a Hungarian doctor, Ignaz Semmelweis, described the importance of handwashing and the sterilization of surgical instruments. Semmelweis worked in obstetrics at the Vienna General Hospital, in a maternity unit that had been set up to address the issue of infanticide – a common practice among underprivileged mothers at that time, mainly because of poverty. The unit took in pregnant women for free, in return for the women being subjects for the training of doctors and midwives. There were two maternity clinics in the hospital, and Semmelweis noticed that in the first clinic there was a maternal mortality rate of 7–15 per cent due to something called puerperal fever, while in the second clinic it was less than 2–7 per cent. This was well known outside the hospital and Semmelweis wrote about desperate women on their knees begging to be allowed into the second clinic. Some of the women gave birth in the streets and were then admitted to the hospital, and he noticed that puerperal fever was rare in these so-called street births.

Semmelweis became deeply troubled by the death rate in the first clinic, saying: 'It made me so miserable that life seemed worthless.' He began examining for differences between the clinics and realized that the only major difference was the people who worked in the clinics. The first clinic was used to instruct medical students while the second was for midwives. This is good science, because it involves careful observation that gives rise to an hypothesis.

In 1847, he figured out what was going on. A friend of his, Jakob Kolletschka, had been accidentally poked with a student's scalpel during an autopsy and died. Semmelweis noticed that Kolletschka had died of puerperal fever, just like the women in the first clinic. He realized that the cause of death was contamination from cadavers – the medical students often carried out dissections and would then work in the first clinic. Semmelweis proposed that the medical students carried 'cadaverous particles' on their hands from their dissections, and so he started a policy of handwashing with chlorinated lime. He chose this because it was best at removing the putrid smell from infected bodies during autopsies. The results were remarkable. A drop of 99 per cent in mortality. He was called the 'saviour of mothers'.

Despite the results, it took time for doctors to accept his instructions. One reason was he couldn't explain why handwashing worked. Many still believed that disease was caused by an imbalance in the four humours and some doctors were annoyed at being accused of having dirty hands when they appeared to be clean.

Semmelweis was critical of these colleagues. He caused such a stir that he was forced to leave Vienna and went to work in Budapest. In 1861 he developed severe depression. He reportedly turned every conversation to the topic of puerperal fever. Criticism of his findings continued, and at one point he wrote a letter to 'all obstetricians' in Europe, criticizing them all. He began drinking heavily and was admitted to an asylum because of his increasingly erratic behaviour, which may have been caused by early onset Alzheimer's disease or even syphilis, which obstetricians sometimes caught from the fluids they picked up from patients. He was beaten by the guards when he tried to escape and died of his wounds – which, ironically, had become infected. He was 47.

The term 'Semmelweis reflex' describes a reflex-like rejection of new knowledge because it contradicts entrenched norms or beliefs. Eminent doctors at the time said: 'Doctors are Gentlemen, and Gentlemen's hands are clean.' Sir Frederick Treves, who was famous for his friendship with Joseph Merrick (also known as the Elephant Man) said: 'There was no object in being clean . . . Indeed, cleanliness was out of place. It was considered to be finicking and affected. An executioner might as well manicure his nails before chopping off a head.' We saw something similar during the COVID-19 pandemic, when it took a while for it to be accepted that it was an airborne disease rather than one that was mainly caught from droplets on contaminated surfaces.

The work of Semmelweis is a great example of how systematic scientific observation can be used to find a solution to a problem. What Semmelweis had uncovered was the role of microbes in infection.

Pasteur's Bean Revolt

John Snow was a surgeon and pioneer in the use of anaesthetics in the 1850s. He was an outspoken sceptic of the miasma theory that claimed that diseases like cholera were spread by 'bad air'. Following an outbreak of cholera in London, he asked residents about their movements and realized that they had all been drawing water at a particular water pump in Broad Street. His analysis was convincing enough to persuade the local authority to remove the hand pump, leading to an immediate decrease in the incidence of cholera.

Snow is known as the father of epidemiology for this work. He also showed that in houses taking water from a section of the Thames polluted with sewage, the rate of cholera was 14 times higher. Snow died relatively young, at the age of 45, from a stroke.

The death knell of the miasma theory happened with the advent of the germ theory of disease. As we saw in chapter 5, what we now know as bacteria were seen by Antonie van Leeuwenhoek in the 1680s using his microscope. He called them 'animalcules', meaning 'little animals'. But that most famous of French scientists, Louis Pasteur, was the one to start the germ revolution.

Like many great scientists, Pasteur was something of a slow starter. He failed his first-year exams at university, finally passing in Dijon with a mediocre grade in chemistry. He eventually obtained a post as professor of chemistry at the University of Strasbourg, marrying the daughter of the rector of the university. No flies on Louis (he must have been very clean). He had five children, but three died in childhood of typhoid, possibly spurring his studies. In 1854 he moved to the University of Lille and began work on fermentation, which he showed was due to microscopic yeast turning sugar into alcohol. Now there's a claim to fame. He also showed how, in contaminated wine, microbes turn sugar into lactic acid, souring the wine. While at Lille he wrote a famous phrase: 'In the field of observation, chance favours only the prepared mind.'

In 1857 he became director of scientific studies at the École Normale Supérieure. He was a disciplinarian and introduced strict rules for the students, leading to 'the bean revolt' after he insisted that the students eat mutton stew with beans (which they hated) every Monday, and 73 out of the 80 students resigned after he banned smoking.

His work on microorganisms in the spoiling of wine and also in milk led him to the idea that microbes might infect animals and humans and cause disease. Pasteur's work vindicated the findings of both Semmelweis and Snow. It also led to Joseph Lister using antiseptics in 1865 to kill microorganisms during surgery.

In the 1870s Pasteur was working on the bacterium that causes anthrax, which was a significant problem in cattle. This had been discovered in 1876 by German bacteriologist Robert Koch, who is often credited alongside Pasteur for the germ theory of disease. Sadly, Pasteur and Koch fell out. This began when Pasteur didn't give Koch sufficient recognition for the discovery of the anthrax bacterium. Not an uncommon problem. Some scientists hate it when their own work isn't cited by a rival. I prefer it if I am cited for work done by someone else, just to annoy them. I always apologize, though. Pasteur's animal work on anthrax was then criticized by Koch, who said that Pasteur had used impure preparations of anthrax and that his work wasn't sufficiently scientific.

Pasteur was hurt and reminded Koch of his groundbreaking work on silkworms. Pasteur had become famous for showing that a disease in silkworms was caused by a microbe, and he saved the entire French silk industry by coming up with a way to detect the microbe in silkworm eggs which could then be discarded. He was also of course famous for coming up with a method for mildly heating milk and wine to preserve them, in a process that became known as pasteurization – of which he also reminded poor Koch, who must have regretted ever saying anything bad about Pasteur. Mind you, Koch may have had a point, as Pasteur's lab notebooks (held in secret by his family until 1964) confirm his genius, but suggested he wasn't always accurate in disclosing his methods. A reviewer once commented that he could be 'unfair, combative, arrogant, unattractive in attitude, inflexible and even dogmatic'. Would he get away with those kinds of behaviours nowadays?

So, Pasteur did away with the miasma theory that many diseases were caused by bad air. But also the idea that life could begin by 'spontaneous generation', which was the term used to describe how life can emerge from non-living matter. Aristotle had proposed it, stating in his *History of Animals* that some animals 'grow spontaneously . . . from putrefying earth or vegetable matter', and it was widely held to be true up until the nineteenth century. Pasteur demonstrated that the growth of microbes in a special nutrient broth wasn't due to spontaneous generation but instead by bacteria getting into the vessel. The evidence involved putting long twisted tubing into the broth (which had been boiled), allowing the microorganisms to get into the broth on dust particles travelling down the tubing. This not only informed the debate on how life starts (proving it doesn't happen spontaneously), but also confirmed the presence of microbes everywhere.

In Germany, Koch had been working away. He discovered not only the bacteria that cause anthrax, but also those responsible for TB and cholera, and as such he is considered one of the founders of bacteriology along with Pasteur. Koch studied medicine at the University of Göttingen and initially researched nutrients and energy, working on the key energy intermediate succinate – a component of the Krebs cycle, important for extracting energy from carbohydrates, that we read about in chapter 5. His wife gave him the gift of a microscope, which he set up in a room off his patient examination room. It was in that room that he discovered the anthrax bacterium. He improved the microscope, was the first to use photography to capture microscopic observations, and was the first to use Petri dishes (named after his assistant, Julius Petri) to grow bacteria. It's somewhat disappointing that his assistant wasn't called Beaker, the assistant of Dr Bunsen Honeydew in *The Muppets*. Koch first demonstrated the Petri dish method of culturing bacteria at a conference in London in 1881. Pasteur was in the audience and exclaimed 'What a great progress, Sir!' They clearly hadn't fallen out at that stage.

The discovery by Koch of the bacterium *Mycobacterium tuberculosis* in 1882 was a major breakthrough, as TB was a huge problem all over Europe at that time, claiming many lives. And here it was, the causative agent. There wasn't much of a reaction to Koch's publication on TB. His mentor, Rudolf Virchow, who was a very eminent physician, was sceptical. He believed all diseases to be caused by faulty biochemistry in cells.

Koch expanded on the discovery in 1884, listing the evidence in the form of what are now known as 'Koch's postulates':

1. The microorganism must be found in organisms suffering from the disease.
2. The microorganism must be isolated from the diseased organism and grown in culture.
3. The cultured microorganism should cause disease when injected into an organism.
4. The microorganism must be re-isolated from the diseased organism and shown to be identical to the original microorganism.

These provide the requirement for proving a microorganism causes a disease, and were important at the time (and still are) because of the disagreement concerning germ theory. Koch later dropped number one, since he found evidence for organisms carrying a microbe and yet not having the disease – in so-called asymptomatic carriers. This was shown to occur in cholera, polio and several other infectious diseases.

Nightingale's Owl

At this point, the more attentive readers may have noticed that the only women mentioned so far in this chapter have been midwives, a victim of domestic violence, suspected witches, and the sufferers of pregnancy-related diseases. It was difficult for women to make an impact in medicine, because they were largely excluded from its scientific study throughout history; though they had always held important roles in healing, with nuns in various religious holy orders founding hospitals. Florence Nightingale stands out as a key person in medicine in the nineteenth century. Her efforts in nursing have a direct link to Semmelweis, Snow and ultimately Pasteur and Koch, in that she was a stickler for hygiene and handwashing.

Nightingale's father had an enlightened view of women's education, and made sure she was taught history, mathematics and philosophy. She displayed what at the time was viewed as an extraordinary ability for a girl in data collection and analysis, which would stand her in good stead later. When she was 18, her father took her off on a tour of Europe and introduced her to Mary Clarke in Paris. She proved to be a great influence on Florence, who wrote in her diary that Mary didn't care about her appearance, and most of all 'was incapable of boring anyone'. Clarke told her that women could be equal to men, and they remained friends for 40 years.

Nightingale had what were thought of as eccentric views and had little respect for upper-class British women who she regarded as 'inconsequential'. When asked if she would prefer to be a woman or a galley slave, she said she would choose the freedom of the galleys.

She decided on a life of helping others, and to the dismay of her mother rejected the notion of marriage and motherhood. She had one persistent suitor, but after nine years told him she wouldn't marry him, for fear of marriage interfering with her calling to nursing. On another trip in Europe, she rescued a baby owl which was being tormented by a group of children. She carried the owl in her pocket for several years, until she went to the Crimean War. In 1853, she took the job of superintendent at the Institute for the Care of Sick Gentlewomen in London.

The following year, she was asked to lead a staff of 38 nurses, along with her aunt Mai Smith (if in doubt, always bring your aunt) to a hospital in Scutari, near the Crimea – which was overcrowded with wounded British soldiers. She found appalling conditions there, with mass infections which were often fatal. She sent a plea to *The Times* for a government solution. The British government commissioned the famous engineer Isambard Kingdom Brunel to design a prefabricated hospital which was shipped to the Dardanelles. Over the course of the Crimean War, Florence's work led to a decrease in the death rate in hospitals from 42 per cent to 2 per cent. Ten times as many deaths happened because of infectious diseases caught on the battlefield or in hospitals than from battle wounds. She knew of Semmelweis's work and implemented handwashing and other hygiene practices. Once she was back in England, she established a training programme for nurses at St Thomas' Hospital, London.

Nightingale had a passion for statistics and the visual presentation of data. Her analysis of the hospital system in Scutari contained statistical charts, which she considered more important than the raw data since politicians would be better able to understand them. She was an early user of the pie chart, and in 1859 she became the first woman elected to the Royal Statistical Society. In 1874 she became an honorary member of the American Statistical Association.

Perhaps most importantly, she wrote the first-ever book on nursing, which was hugely influential – the aptly named *Notes on Nursing*. It was written at a time when simple rules about health were only just beginning to be understood. It contained advice on ventilation, nutrition, bedding, hygiene (this was a major focus – of the nurse, patient and environment), and what she called 'chattering hopes and advices', by which she meant the false assurances and recommendations of family and friends to the sick. The book is still in print, and its message is as relevant today as when it first appeared in 1859.

Avoid 'Chattering Hopes and Advices'

The first vaccine was deployed by Edward Jenner for smallpox in 1796. Folk medicine had described how if someone was infected with cowpox, they wouldn't succumb to smallpox. Jenner knew about the work of Lady Mary Wortley Montagu, who had brought a technique called variolation to England from Turkey. The Turks had been using fluid from smallpox lesions to protect against reinfection, and Montagu introduced this to England in 1721 by testing it on condemned prisoners in Newgate prison, although she gave it to her own children first. They all survived and were released.

Cowpox was shown to be as effective at protecting from smallpox by Jenner and also safer, since the Turkish method sometimes actually gave the person smallpox. The method was widely adopted, preventing smallpox – which had been a scourge on humanity, killing one in three who became infected and disfiguring many who survived. Napoleon was so impressed that he said he could 'refuse this man nothing', even though France and England were at war.

Louis Pasteur also followed Jenner's work closely. He had figured out that cowpox was somehow a weakened form of smallpox, and so he made weakened forms of the bacteria that caused cholera and anthrax and the virus that caused rabies. He chose the word 'vaccine' in honour of Jenner, from the Latin for 'cow'.

From these practical observations and advances, the inner workings of the immune system began to be unravelled, with immunologists demonstrating that vaccines enlist specific cells called T and B cells (sub-populations of white blood cells) to fight the infection, with some of them becoming memory cells – at the ready should infection strike again. This T and B cell response happens in natural infections too, it's just that vaccines are safer as they don't cause disease. Many vaccines have emerged in the 20th and 21st centuries, preventing major causes of disease and death – notably influenza, pneumonia, measles, diphtheria, hepatitis, polio, cervical cancer (which is caused by human papilloma virus) and most recently of course the success of the vaccines for COVID-19. In terms of lives saved, the development of vaccines can be considered the biggest contribution in medicine.

Sadly, we saw an awful lot of anti-science during the COVID-19 pandemic, with objections to vaccines and masks that weren't based on sound evidence, and advocacy for treatments that were unproven and even downright dangerous, like hydroxychloroquine and Ivermectin. In the US, the death toll has exceeded that of the 1918 pandemic despite 100 years of scientific advances. Surely all that scientific knowledge that has accumulated since the 1918 pandemic could have been used to save people from COVID-19? Why it didn't remains a mystery to many of us, but is partly to do with politics. A survey carried out by the Pew Research Center revealed that Democrats were less sceptical of the COVID-19 vaccines than Republicans. Vaccine uptake was also much lower among Republicans, with a study in December 2022 finding that in the US 37 per cent of Republicans remain unvaccinated compared with 9 per cent of Democrats. This is likely to be the main reason why excess deaths are higher among Republicans than Democrats – as much as 76 per cent higher in two states that were examined, Ohio and Florida. This isn't to say that being Republican causes you to be wary of vaccines – that would be falling into the trap of confusing correlation and causation – but perhaps a greater sense of individualism and distrust of institutions makes a person more likely to both vote Republican and be a vaccine sceptic.

Ivermectin is a particularly salutary tale. It is a drug used to treat infestation with parasitic worms, and was proposed as a drug to kill the SARS-CoV-2 virus but with no clear mechanism. Some studies were published claiming an antiviral effect on cells infected with virus in the laboratory, but the drug concentrations needed could not be achieved in patients. Online misinformation and advocacy campaigns promoted the use of Ivermectin. Conspiracy theories circulated about governments and scientists suppressing evidence that Ivermectin worked against COVID-19, to allow pharmaceutical companies to make money from the vaccines. Many studies were flawed, with some even accused of containing falsified data. Some were retracted because of the flaws, which is of course a good thing and how science should work. The large double-blind randomized controlled trial (RCT) ACTIV-6, which was published in October 2022, showed no benefit when comparing patients treated with Ivermectin to a placebo. It has been called the 'final nail in the coffin' for Ivermectin as a therapy for COVID-19.

An RCT involves the random allocation of participants into either a group that receives the treatment (in this case Ivermectin) or a group that receives a placebo. Neither the participant nor the scientist knows who has received the treatment or the placebo. This is why it is called 'double blind'; it removes any biases in the scientist or participant.

The Ivermectin saga illustrates the importance of regulation and high standards. Properly done science proved itself as the most important ally against COVID-19, despite what you might read when going down the rabbit hole of social media. The trials that showed that the main vaccines for COVID-19 that were approved (the Pfizer/BioNTech, Moderna, AstraZeneca and Janssen vaccines) were all RCTs, and showed that the vaccines were safe and could protect against disease. The results from the trials were borne out in the use of these vaccines in billions of people. Importantly, any risks from the vaccines (and there are some, as with all medicines) are much less than the risks from being infected.

As an immunologist and COVID-19 researcher (my lab works on therapies for COVID-19), I was interviewed a lot in the media – in Ireland but also in the UK and US. As an immunologist I strongly advocated for vaccination, which resulted in me being trolled on social media. The Ivermectin advocates also came after me after I questioned its use. A survey of scientists carried out by *Nature* magazine in October 2021 revealed that 60 per cent of scientists commenting on COVID-19 in the media had their credibility attacked, 40 per cent suffered emotional distress because of attacks and 30 per cent suffered reputational damage. Some were assaulted or received death threats. These things happened to me too, but I never wavered. I saw it as my job to keep the public informed – many of whom were frightened, especially if they were older, about the greatest health crisis to hit the world in 100 years. I stuck to the mantra of 'Keep Calm and Trust the Science' – and Florence Nightingale's advice to avoid 'chattering hopes and advices'.

Fleming's Filthy Lab

Apart from vaccines, the other great advance in twentieth-century medicine was the discovery of antibiotics by Alexander Fleming. Penicillin was found accidentally by Fleming in 1928, when his sloppiness in the lab led to a bacterial culture being contaminated with a mould, which killed the bacteria. It is ironic that both Pasteur and Fleming had filthy labs, since their stellar contributions to human health have been in the area of infectious diseases.

Penicillin was isolated from the mould through a process of trial and error, and went on to save countless millions of lives. It works by destroying the cell wall of bacteria and so they burst open and die. It stops the production of a substance called peptidoglycan, which is a key building block for the cell wall of many bacteria. We now know that antibiotics are made naturally by bacteria themselves, as well as fungi. This gives them a growth advantage, as they can kill off the competition for space and nutrients. They are a natural weapon used in the fight for life at a microscopic level, and we use that weapon ourselves to kill nasty bacteria. Thanks, nature.

Many other antibiotics as well as antiviral and antifungal drugs have now been discovered. Unfortunately – as we saw with COVID-19 as it mutated through the Greek alphabet from the Alpha variant to the Omicron variant, and with the bacterium MRSA (which stands for methicillin-resistant Staphylococcus aureus, methicillin being an antibiotic) – the microorganisms will fight back against us. The variants in COVID-19 made them more resistant to our own immune system, which means booster shots are needed, while in bacteria they become resistant to antibiotics. Resistance, driven by selective pressure from a vaccine or antibiotics, is a fact of life.

Fortunately, we now have very powerful and rapid diagnostic tools that identify exactly how resistance is occurring, which allows for the selection or design of new treatments or vaccines. We knew almost immediately exactly which parts of the COVID-19 spike protein had evolved to give us the Omicron variant, because of rapid genome sequencing techniques. Our immune systems make antibodies to recognize the spike protein, so if that changes, those antibodies become less effective. But thanks to RNA technology, where an RNA molecule can be rapidly made for the latest variant in the spike protein, a specific vaccine became available within months that boosted protection – hence the term 'booster'. Koch would have been impressed, and even Pasteur would have agreed with him.

What Will You Die Of?

The advances in many areas of biology have allowed a whole range of new medicines to be discovered, to treat many diseases – including the old established favourites such as arthritis, asthma and cancer, the Western lifestyle problems of obesity and heart disease, and the problems of ageing we have created through our extended lifespan. All the knowledge about how the body works can be deployed in the hunt for new drugs. The use of statins to lower cholesterol and prevent heart disease has been a big contributor to increased average life expectancy in Western countries. Statins work by targeting an enzyme called HMG-CoA reductase that is key in making cholesterol. Blocking it lowers the amount of cholesterol in the body, and because cholesterol is the main ingredient that clogs up blood vessels in the heart, lowering it reduces the risk of a heart attack.

Other reasons for us living to a ripe old age in good health include better treatments for cancer and for inflammatory diseases like rheumatoid arthritis and inflammatory bowel disease, which target the out-of-control immune system that lies at the heart of autoimmune diseases. There is great optimism that lung cancer, pancreatic cancer, bowel cancer and breast cancer will no longer kill us but will be manageable. In 2022, Joe Biden set the goal of reducing the cancer death rate by 50 per cent within 25 years. That progress may well come from getting your own immune system to kill the cancer cells. Evidence for this comes from melanoma, a form of skin cancer that can be caused by excessive exposure to the sun. Over the last 10 years, new drugs were introduced – referred to as checkpoint inhibitors – that enhance the immune response against tumours. Before this, patients whose tumours had metastasized (spread to different parts of the body) had a survival rate of 5 per cent five years after diagnosis. The new drugs have increased this to over 50 per cent. Science did this!

Better imaging technologies (for example, the invention of the magnetic resonance imaging or MRI scan, which can look inside your body in great detail to find out what's going wrong) and molecular diagnostics (such as PCR, familiar from COVID-19 testing) have allowed for more accurate diagnoses.

The genetic basis for diseases is also a very active area of research, since so many diseases happen because of differences in genes, often combined with particular lifestyles. In a project called Our Future Health, 5 million adults in the UK are having their DNA analysed, and any differences linked to lifestyle and health recorded, in an effort to find genetic variations that make people more likely to develop chronic diseases such as heart disease and type 2 diabetes. This in turn could lead to new therapies that target the genetic differences.

Advances in gene therapy and editing – using technologies like CRISPR, which we read about in chapter 6 – as well as stem cell therapy raise the prospect of many diseases being treatable or completely preventable in the future. Cell therapy involves taking cells out of a patient's body, modifying them genetically to carry out a specific function, and then putting them back in. It's especially effective in a treatment called CAR-T, where T cells are engineered to kill leukaemic cells, offering a cure for certain types of leukaemia without having to resort to bone marrow transplantation. It might even be possible to grow brand-new organs from stem cells to replace organs as they wear out, and therefore defy ageing.

There is still a pressing need for treatments involving diseases of the brain, including age-related neurodegenerative diseases like Alzheimer's and Parkinson's disease, and mental health disorders such as depression and schizophrenia, which remain difficult to treat because we don't know enough about what is going wrong in the brain in those diseases.

As we continue to learn more and more about the inner workings of our bodies in health and disease, science will come up with all kinds of novel approaches to treat the diseases that afflict us. Science might even come up with a way to slow down the ageing process. Even billionaires such as Jeff Bezos are starting to invest in anti-ageing companies, so space is no longer the final frontier – it's time itself. You might wonder what you might die of, given all these advances. Boredom? Not if you read my books of course, and they are also helpful on those long space flights.

Science made medicine respectable, and should be the only method to use when it comes to discovering, assessing and deploying new treatments. That's not to say that the 'caring' component of medicine isn't important. When we are sick, we crave comfort and reassurance, and it is good for us. The placebo effect, where belief in something can have positive effects, is real. That's how lots of alternative therapies 'work'. But a placebo won't cure cancer, or treat dementia, COVID-19 or schizophrenia. Science will.

8. Where Is My Mind?

Progress has been very slow in the effort to prevent or treat diseases that afflict the mind. These present a significant burden on humanity, and include such conditions as depression and schizophrenia, as well as neurodegenerative disorders like Parkinson's and Alzheimer's disease. All of us know someone with such diseases, and wouldn't it be marvellous if doctors could offer a better solution for teenagers with eating disorders or middle-aged people with severe depression or older people who are losing their memories?

One reason why there haven't been many advances in therapies is the mind remains one of the biggest mysteries in all of science. We think it lives in the 1.2 kilograms of pink/white, soft-tofu-consistency brain tissue we protect in our hard bony skulls. But we still have a limited understanding of how it works.

We have no real idea how our brain can incorporate all the control systems that regulate our physical activities alongside encapsulating our sense of self, our consciousness, our sentience – the essence of being human. How could we possibly understand it? It's all so complicated. How could the squishy pink mass of wet tissue in our skulls do all these things? I mean, it's only made of flesh and bone, right? And yet it does. It's the most complicated wet lump in the universe.

It's like the knowledge our Stone Age ancestors had about the sun, the moon and the stars. They could observe some of the component parts but had no understanding of what the universe comprised, how it worked or how it all hung together. But, thanks to some of the concepts and technologies developed by the human minds that feature in this chapter, I think we are on the brink of making major advances in understanding our minds. Finally, we will be able to answer the question 'How is it we can think about how we think?'

Of course, all throughout history we've mused about it. As ever, it starts with those brainy sods the Greeks. They certainly sat around a lot in the sun, drinking wine and doing an awful lot of thinking. Maybe Monty Python got it right when they sang about Socrates, Plato and Aristotle being drunk all the time. But all that philosophical contemplation about the mind didn't really get them very far. They couldn't make their minds up as to whether the brain or the heart was the seat of intelligence. If only they had stayed sober, they might have figured it out.

The Egyptians before them had concluded that the heart was where the action was. They tugged the brain out of the nose during mummification and threw it away, because they felt it wasn't especially important. Aristotle thought the function of the brain was to cool the blood coming from the heart, but then Hippocrates yet again reared his clever head. He was of the opinion that the brain was the place where intelligence lay. For evidence, he noted that most of our sense organs – the eyes, the ears and the tongue – are in the head, near the brain. So, he reasoned, intelligence must also be there. Hippocrates left us with a remarkably prescient quote on what he thought the brain was for: 'And men ought to know that from nothing else but the brain come joys, delights, laughter and sports, and sorrows, griefs, despondency, and lamentations.' Greek anatomists revealed the complexities of the structure of the brain by dissecting it. They showed that there were different structures within the brain – notably the cerebrum and the cerebellum, provoking further musings on what these brain parts might do.

As we read in the previous chapter, Galen enjoyed dissecting animals, and dissected the brains of cows, Barbary apes (presumably without looking at their faces) and pigs. He thought the cerebellum, which is the toughest part of the brain, somehow controlled muscles in the body. He figured out that the spinal cord connected to the brain and to muscles. He also noted that patients lost their mental skills when they had a brain injury.

Arab physician Al-Zahrawi in the tenth century was a pioneering neurosurgeon, who performed surgical treatments for head injuries. He was the first to tie off the temporal artery to treat severe migraine, 600 years before a French neurosurgeon called Ambroise Paré claimed to be the first to do this procedure. Though Paré deserves some credit, as he did it to his own temporal artery while fully awake, such was his desperation.

Al-Zahrawi must have hypothesized that too much blood flowing into the brain might be a cause of migraine. He wasn't too far off the mark, as the latest treatment for migraine involves stopping a protein called CGRP that regulates blood flow in the brain. In people who suffer from migraines, for unknown reasons too much CGRP is made, and this causes the vasodilation of blood vessels. That in turn can increase the sensation of pain, and so dialling down the effect of CGRP brings benefits. When I was a PhD student in London one of my fellow students, Sue Brain, now an eminent professor at King's College London, was working on CGRP and how it causes vasodilation. At that time, she was mainly interested in inflammatory diseases like arthritis, which also feature increased blood flow – in that case into the joints, causing pain and inflammation. Sue didn't work on migraines then, although she has since made important contributions to figuring out CGRP's role in them. This is wholly appropriate, given her surname.

Skull Blasting!

René Descartes was the first to discuss the issue of how the brain relates to the mind. He was a French philosopher and scientist, now perhaps best known for the phrase '*cogito, ergo sum*' – 'I think, therefore I am'. He was drawn to science from reading the works of Galileo, another example of how a brilliant scientist can impact science in an even bigger way than simply their own research.

In 1619, Descartes shut himself away in his room and reported that he had three dreams. He said that in the dreams a 'divine spirit' revealed a whole new philosophy to him. He's thought to have had an attack of exploding head syndrome, a great name for a condition if ever there was one. It's a sleep disorder where you hear a loud noise in your head, like a gunshot or fireworks, as you fall asleep, although no such noises have happened. The sensation is sometimes accompanied by muscle jerks. It's also known as episodic cranial sensory shocks, and the muscle movements are called myoclonic jerks. You can't beat doctors and scientists for complicating things with fancy names. Reassuringly, although alarming, exploding head syndrome is harmless and doesn't indicate anything serious. Descartes' famous phrase came out of this experience. Maybe it should have been 'I think I've heard a loud noise, therefore I am'. The visions inspired him to be a scientist, and by the 1640s he had become one of Europe's leading scientists and philosophers.

In 1649 he was invited by Queen Christina of Sweden to set up a new scientific academy and tutor her in his ideas about, of all things, love. She was interested in a book he'd written about emotion called *The Passions of the Soul.* The book came about because of correspondence between Descartes and Princess Elisabeth of Bohemia. She had asked him questions about happiness and the passions that afflict us – love, hate, sorrow – and Descartes not only responded but decided to write a book on the subject. When it came to his own passions, he wrote: 'As a child I was in love with a girl of my own age, who was slightly cross-eyed. The imprint made on my brain by the wayward eyes became so mingled with whatever else had aroused in me the feeling of love that for years afterwards when I saw a cross-eyed woman, I was more prone to love her than any other, simply for that flaw, all the while not knowing there was the reason.' Whatever turns you on.

Descartes wrote that he wanted to explain passions not as a philosopher, but as a physicist. In *The Passions of the Soul* he explored the relationship between the mind and body, known as mind–body dualism. The body was considered a physical entity, while the mind (or soul) was something mysterious. Descartes defined emotion as 'perceptions, sensations or commotions' of the soul that were caused by some 'movement of the spirits'. By 'spirits', he meant something produced by the blood that could 'move the body'. He concluded that the pineal gland was where the mind and body interacted. He was effectively saying that the mind had a physical basis, which was a radical idea in the seventeenth century.

Queen Christina was clearly jealous of Princess Elisabeth's correspondence with Descartes (it must have been great to have royal women fighting over him). He became her tutor largely for money, as he was poor due to bad business decisions, which involved giving her lessons at 5 a.m., three mornings a week in her draughty castle. It's no wonder they fell out. She disagreed with his 'mechanical philosophy', which argued that the universe was one big machine to be understood like any other machine, and he was bored by her interest in ancient Greek. He is thought to have caught pneumonia during these lessons and died. However, there is also some evidence that he was poisoned. Not by Queen Christina (although it sounds like she wanted nothing more), but by a Catholic missionary in her retinue, Jacques Viogue, who disagreed with his views on religion. Viogue was working in Sweden and is alleged to have given Descartes an arsenic-laden communion wafer. He was apparently worried that Descartes, with his teachings on metaphysics, might thwart his efforts to convert Queen Christina to Catholicism.

As a Catholic in a Protestant country, Descartes was buried in a graveyard used for orphans. His remains were finally taken to France and buried in Saint-Étienne-du-Mont, where they stayed until the early nineteenth century. In 1792, the National Convention – the first parliament of the French First Republic – recommended transferring his remains to the Panthéon in Paris, but he was finally reburied in the Abbey of Saint-Germain-des-Prés in 1819. Before being reinterred, the coffin was opened, and to their surprise, the onlookers noticed there was no skull. Descartes had been a victim of what was called 'skull blasting': the practice of removing the heads of prominent people (after they had died, obviously enough) and then selling the whole skull or parts of it to collectors.

One of those present for this ghastly discovery was the pioneering chemist Jöns Jakob Berzelius, who was visiting from Sweden. Quite why he wanted to look inside Descartes' coffin remains unclear. He somehow learned that the skull had been sold to a casino mogul, and he convinced the man to sell it to him.

Berzelius was a firm believer in empirical evidence, and so he must have believed that the skull he bought was indeed that of Descartes. Someone had written across the forehead: 'The skull of Descartes, taken by J. Fr. Planstrom'. Sounds legit. An historian hired by France's Académie des Sciences was given the task of piecing together what had happened. He found that Planström, one of the guards charged with protecting the remains while they were awaiting transport to France, had indeed taken the skull. When Planström died, he owed money to a Stockholm brewer who was given the skull as payment. When the brewer died, it passed on to his son and then to a long line of collectors, some of whom wrote their names on the skull. It eventually found its way to a scientist called Anders Sparrman, whose collection was auctioned off after his death, and ended up with the casino owner. Berzelius bought it and returned it to France, where it has resided for 200 years. You can visit it in the Musée de l'Homme in Paris.

For a man who wrote about mind–body dualism, it is fitting that Descartes' skull was finally reunited with his body. But four centuries later, precisely how the mind relates to the brain is still not clear.

Nobody Laughs and the Frog Dies

Our next advance also has a kind-of French connection, since it involves frogs' legs. It was made by Luigi Galvani, an Italian doctor. In 1780, he and his wife, Lucia, observed that muscles from dead frogs' legs twitched when struck by an electrical spark. But how on earth did they come to do that experiment? Luigi had studied medicine at the University of Bologna, reading the works of our old friends Hippocrates, Galen and Avicenna, and he had also studied surgery. Lucia was one of his professors' daughters, and having married her, after graduation Galvani got a job at the university as a lecturer in surgery and anatomy. Smart move. Lucia and Luigi would work very closely together, and she helped write her husband's publications. Typical of the times, Lucia got no recognition for her contributions. Luigi, meanwhile, became a member of the highly prestigious Benedictine Academy of Sciences, and so was obliged to publish one research paper every year, which he did until his death. However, he got fed up with this publication schedule, as there were often delays. One year, he worked up a series of observations on hearing in birds and humans, but was plagiarized by another Academy member – one Antonio Scarpa. The 'publish or perish' pressure existed then just as it does today. It's the bane of the life of academic scientists, as without publications they can't get grants or promotions.

This led to Galvani giving up on that area of science entirely, and instead he moved to what was called 'medical electricity', which involved applying electrical stimuli to the human body. The use of electricity in medicine goes back to

46 BCE, when the Roman physician Scribonius Largus (who was surely in the Harry Potter books) recommended that his patients stand on a live black torpedo fish to relieve the pain of gout. Sounds desperate, right? But then, gout can be extremely painful. Other Roman physicians, including Galen, recommended electric fish for the treatment of headache pain and a prolapsed anus – I would love to see that prescription: 'Take a large electrical fish and shove it . . .' Eye-watering! Then, in the 1700s, doctors finally moved away from fish and began using devices that had been invented to generate electricity. John Wesley, an English clergyman who founded the Methodist movement, promoted electrical treatment as a panacea for all ailments. Devices with names like the 'Oudin coil' and 'Pulvermacher's chain' were used to treat patients with different ailments, mainly those involving pain like arthritis and back pain.

Galvani made his discovery with frogs' legs accidentally. He was slowly skinning a frog. Why this was being done slowly, or at all, remains a mystery, but as British comedian Barry Cryer has said about analysing jokes: 'It's like dissecting a frog. Nobody laughs and the frog dies.' Galvani and Lucia were at a table where they had previously created static electricity by rubbing frog skin with a metal scalpel. His assistant touched the skinned leg of the frog with the scalpel, sparks flew, and the frog's leg kicked as if the frog were alive.

We now know that the electric current had stimulated the sciatic nerve, which transmitted a signal to the muscle in the leg, causing it to contract. Galvani coined the term 'animal electricity' to describe the forces acting on the muscle.

Galvani's work influenced another Italian investigator working on electricity, Alessandro Volta. Born in 1745 in the northern Italian town of Como, in 1779 he was appointed professor of physics at the University of Pavia. His lectures there on electricity were so popular that no less a figure than the Holy Roman Emperor Joseph II ordered the construction of a new lecture theatre, the Aula Volta. The emperor also provided him with substantial funding for his research.

In 1800 there was an argument between Galvani and Volta. Galvani concluded that electricity was intrinsic to the frog's leg, whereas Volta concluded that the movement in the muscle depended on the metal cable Galvani had devised to link the charged scalpel to the muscle. Galvani said that the animal electricity came from the muscle in the pelvis of the frog. Volta concluded that it was a metallic electricity that occurred because of different metals in the scalpel and cable. It was this dispute that led Volta to make the first-ever electric battery, known as the voltaic pile.

He did this by showing that the best way to produce electricity was with two different metals. He first used wine goblets filled with brine, into which he dipped zinc and copper electrodes. He then replaced the goblets with the voltaic pile, comprising alternating copper and zinc strips separated by a layer of cloth or cardboard, bathed in either saltwater brine or sulphuric acid (which are electrolytes). This provided a stable electrical current flowing from copper to zinc. The unit of electric potential is named in his honour as the volt. Volta was so well regarded in Europe that he was made a count by Napoleon Bonaparte in 1810.

Galvani's name stays with us in the term 'galvanism', which Volta coined to describe how a chemical action can cause an electrical signal. We also have 'galvanize', meaning to shock someone into taking action. Volta made the battery to disprove Galvani's ideas, but his battery had various flaws that meant it was never used as a source of power – most notably leakage of the electrolyte, causing the battery to short-circuit and a short battery life of less than one hour. However, improvements were made, leading to the batteries we are all now so familiar with and that power mobile phones and electric cars and the very computer I am writing this book for you on. These advances wouldn't have been possible without the work of Volta – or at least would have taken longer. The story of Galvani and Volta shows how arguments between scientists can not only galvanize them into action but make the world better.

Galvani's work also inspired Mary Shelley to write *Frankenstein*. At the age of 17 she was on a Grand Tour of Europe. This was effectively a gap year for upper-class young European men when they passed the age of 21, and was seen as an educational rite of passage. The idea was to expose young men to the Renaissance culture of Europe but also Greek and Roman history. At the time it was an unusual thing for a woman to undertake.

As part of the tour Mary stayed in the German town of Gernsheim, 17 kilometres from Frankenstein Castle. She had read that, 200 years prior, there had been an alchemist in the castle who carried out experiments that frightened the locals. Her travelling companion was Percy Bysshe Shelley, the famous poet who she had married. Percy told her about Galvani's frog experiment. Or more likely, Byron's personal physician who was with them, the brilliant 20-year-old John William Polidori, regaled them with reports of the latest developments in medical science. They reached Lake Geneva and met up with Lord Byron, another famous and tragic romantic poet who we will meet again later. Stuck indoors because of bad weather, they decided to amuse themselves with a competition to see who could write the best horror story. The bad weather was due to a volcano, Mount Tambora, that had erupted the year before and created the so-called Year Without a Summer. The story Mary wrote was *Frankenstein.* She was only 18 years of age at the time, and the book was published two years later. In *Frankenstein*, Victor Frankenstein creates life from raw materials supplied by 'the dissecting room and the slaughter-house' and using electricity, but he is horrified by what he has made. Because it used current scientific ideas to shock the reader, *Frankenstein* is recognized as the first true science-fiction horror story. I wonder what a group of young people trapped inside by a storm would do nowadays. Literature's loss is TikTok's gain.

Frankenstein describes a scientist deciding to use their laboratory to carry out a scientific experiment which goes badly wrong. How many times have we read that scenario? *Spider-Man*? *The Incredible Hulk*? These stories owe a lot to *Frankenstein.* And this classic work mixing art and science has also proven to be very adaptable. The first theatrical version opened in London in 1823 to large audiences and publicity: 'Do not take your wives, do not take your daughters, do not take your families.' Mary Shelley herself attended and noted: 'In early performances all the ladies fainted.' The horror genre begins!

A Hole in Phineas's Head

As the 1800s progressed, anatomy as a science continued to advance in medical schools, and anatomists began dissecting the brain more and more and could ascribe functions to different parts. French physician Jean Pierre Flourens was important here. In Paris in 1815 he started a series of experiments in which he created lesions in specific parts of the brains of living rabbits and pigeons, and then he carefully observed the effects on muscle movement, consciousness and behaviour. He was investigating what at the time was called localizationism. Yet again, you can't beat scientists for coming up with obscure terms for things.

Localizationism theorized that different parts of the brain govern different functions. When Flourens removed the cerebral hemispheres, meaning the two halves of the part of the brain called the cerebrum, all senses were abolished. Removing the cerebellum, the part of the brain at the back of the skull, abolished balance and muscle coordination. Removing the brainstem caused death. He concluded that the cerebral hemispheres were responsible for higher-order functions of the brain (effectively, the thinking activities), the cerebellum coordinated movement, and the brainstem controlled vital functions such as breathing and circulation. He couldn't find the parts responsible for memory and concluded the entire brain was somehow needed for that. Remember, this was over 200 years ago and all he had was a knife and some rabbits.

Flourens was highly regarded in France and was elected to the French Academy instead of novelist Victor Hugo – which must have made Hugo, the author of *Les Misérables*, tres misérables. Somewhat unfortunately for Flourens and his scientific legacy, he vociferously criticized Darwin and his theory of natural selection, but he saved face in the English-speaking world at least, as his book on 'the fixity of species' was never translated into English. This tells us something important about the scientific process. To be a scientist is to be sceptical, and in many cases individual scientists will question the validity of the work of others. Flourens did brilliant work on brain regions and he also promoted the use of chloroform as an anaesthetic during medical procedures. Both contributions were found by other scientists to be valid and have been incredibly useful. But his rejection of Darwin's theories and his promotion of creationism were not supported by science and have been rejected. Science iteratively evaluates the views and work of individual scientists, and sooner or later comes to a consensus view.

Further insights into the functions of different parts of the brain came in 1848, when John Martyn Harlow described a patient called Phineas Gage. Phineas was employed as a foreman working on the construction of the Hudson River railroad in New York state. He was an expert on explosives. His job involved boring holes into rock, putting explosives into the hole and then adding sand to contain the blast's energy, directing it into the surrounding rock. This involved the use of a tamping iron to pack the sand into the hole. One day, while Phineas was tamping, the tamping iron sparked against the rock, setting off the explosives. It rocketed from the hole and entered the left side of Gage's face, passed behind his left eye and then exited from the top of his skull. A rare event indeed. The tamping iron landed some 25 metres away. A witness said it was 'smeared with blood and brain'.

Gage fell onto his back and had a convulsion, but quickly recovered. He could walk and speak and got onto his ox cart, which took him back to his lodgings. A local doctor found him sitting on a chair outside his hotel, and said he was addressed by Gage in 'one of the great understatements of medical history': 'Doctor, here is business enough for you.' The doctor saw the open wound and 'pulsations of the brain'. He reported how Gage got up and vomited, with 'about half a teacupful of the brain [exiting the hole], falling on the floor'. The doctor took Gage to the more experienced Harlow, who cleaned up the wound and bandaged Gage's head.

This fortuitous accident for neuroscience (but not for poor Phineas) revealed how the brain was involved in personality. His frontal lobe, a part of the brain just behind the forehead, had been damaged by the iron rod, and his injury indicated that part of the brain is involved in personality. After his injury his friends said he was 'no longer Gage'. He made a gradual recovery but the changes in his personality were attributed to the brain damage he had suffered. Before the accident, Gage was described as hard-working, responsible and well-liked. However, after the accident his former employers refused to take him on again, saying that he was now 'fitful, irreverent . . . manifesting but little deference to his fellows . . . obstinate'.

He became famous, making public appearances, including at Barnum's American Museum in New York, which was popular with people wanting to see what were then termed freaks of nature. A limerick was even written about him:

A moral man Phineas Gage
Tamping powder down holes for his wage
Blew his special-made probe
Through his left frontal lobe
Now he drinks, swears and flies in a rage

He eventually emigrated to Chile, took a job as a stagecoach driver and lived a fairly normal life. Gradually he returned somewhat to his old self. His structured life in Chile was said to help with that. Following several seizures, he died aged 36. His case informed the rehabilitation of others with frontal lobe injury, and so some good came of Phineas's unfortunate accident. And the link between brain and personality had been established.

Harlow eventually got possession of Gage's skull from his family. With permission – not by skull blasting. He also obtained the tamping iron, which Gage had always kept with him as a sort of lucky charm. The iron is now in a museum with the following inscription: 'This is the bar that was shot through the head of Mr Phineas P. Gage.'

Silvery Brains

There was still the issue of how the brain actually worked. A particularly knotty question was how the brain might link to muscle movement. As we saw with the frog experiment, electrical activity had been shown to be involved somehow. In 1850, German physiologist Hermann von Helmholtz even figured out how fast an electrical impulse could pass down a nerve, using the sciatic nerve from, you guessed it, frogs – but also calves. The speed was calculated to be in the range of 24.6–38.4 metres per second. That's a speed of around 100 kilometres per hour, so as fast as a car.

Camillo Golgi made the next important discovery. Golgi studied medicine at the University of Pavia under Cesare Lombroso, an Italian criminologist, who believed that a man's face could predict the crimes he might commit – which had no scientific basis whatsoever. Golgi preferred to work in the lab rather than, say, trying to correlate a man's features with any criminal tendencies. In 1872 he took up a post that allowed him to set up a lab in a refurbished hospital kitchen.

When he started his experiments, the brain was hard to

visualize in any kind of detail. It's not clear how Golgi hit on the dye silver chromate, but it allowed distinct cells – neurons – to be seen in brain samples. He wrote to a friend about his discovery: 'I am delighted that I have found a new reaction to demonstrate, even to the blind, the structure of the interstitial stroma of the cerebral cortex.' Quite what he meant isn't known. Sounds like he was exaggerating to his friend.

Golgi concluded that the brain was a huge network of very fine fibres (which we now call axons) emanating from the central part of each neuron (which we now call the cell body). His dye method revealed the intricacies of neurons throughout the brain. He was the first to describe the fine structure of such brain parts as the cerebellum, hippocampus (so named because it looks a bit like a seahorse, which belongs in the *Hippocampus* genus) and spinal cord. His staining technique was so sensitive that he also saw structures inside cells, which are named after him: the Golgi apparatus. These structures were only accepted by the scientific community 50 years later, when observed directly with the hugely powerful electron microscope.

Although Golgi was the first to observe the dense fibres that make up the brain, it was another neuroanatomist, Santiago Ramón y Cajal, who in the 1890s was the first to conclude that the brain was made up of billions of interconnected neurons (he used an improved version of Golgi's silver stain method). Ramón y Cajal had been rebellious as a child with an anti-authoritarian attitude. His father, an eminent anatomist, apprenticed him to a shoemaker and then a barber. But he also took him to graveyards to help him retrieve human remains for anatomy work, and it looks like it was this rather gruesome activity that inspired him to become an anatomist himself. Ramón y Cajal spotted that, although the brain was made up of neurons, there were gaps between them – named 'synapses' in 1906 by British scientist Charles Scott Sherrington – which would become a very important finding. Ramón y Cajal also hypothesized that the ability of neurons to grow in humans and create new connections was the basis for learning. This was the first attempt to explain this key function of the brain.

Increasingly, scientists began to examine how neurons worked. Studies in what became known as electrophysiology revealed the interconnectedness of neurons and how they communicated with each other. It was known to involve the transmission of an electrical impulse along nerve cells – work that goes all the way back to Galvani and his frogs' legs – but there was the problem of the synapse: the nerves didn't touch each other directly, so how did an electrical signal bridge the gaps? The brain at this stage was seen as a dense network of cells with long tubes coming out of the central part of each neuron, somehow connecting to other tubes across gaps (synapses). And there were billions and billions of them. It was no wonder that scientists were clueless as to how such a complex arrangement might work, let alone be able to explain something like consciousness.

Hangxiety

In 1914, Henry Dale described the first neurotransmitter, acetylcholine, which acts as a communication molecule between neurons. But it was very difficult to study the electrical activity of neurons, mainly because of their small size. Two British scientists, Andrew Huxley and Alan Hodgkin, therefore turned to the enormous nerve from the longfin inshore squid, the squid giant axon. It is around 1.5 millimetres in diameter and about 1 metre in length. Like a long string of spaghetti.

Huxley and Hodgkin went to the Marine Biological Association laboratory in Plymouth to isolate the axon. But even this was difficult, because a nerve impulse down the axon lasts only milliseconds, consistent with the speeds reported earlier by Helmholtz. Huxley and Hodgkin could eventually record the currents surging down the axon, calling the signal an action potential, which is a change in electrical potential associated with an impulse moving along the membrane of a neuron or muscle cell. Then the Second World War intervened and both men were sent to work on radar. Six years later, they were back at work in Plymouth and they figured out that the action potential running down the squid axon was happening down the outer membrane of the neuron. Electrically charged sodium ions were shown to diffuse inwards across the membrane, while potassium ions diffused out, all down the neuron. This is somewhat like a domino effect, with dominoes falling but standing up again as the signal passes from one domino to another. The publication of their findings in 1952 was a breakthrough moment – they are the basis for most neuroscience analysis of neurons since and has been key to the development of new drugs for conditions like depression, stroke, pain, epilepsy and nausea.

But how does the action potential link to neurotransmitters? When the action potential reaches the end of a neuron it causes the release of acetylcholine, which rapidly diffuses across the synapse to the next neuron, which then triggers another action potential propagating the signal further down the nerve. It's like the last domino pushes a ball, which rolls across a gap to trigger the next domino cascade. That ball is a neurotransmitter. These findings in effect explain in broad terms how neurons communicate with one another in the brain.

Neurotransmitters – signalling molecules that are released by a neuron and affect another neuron across a synapse – became a focus for neuroscientists, who believed they would reveal the innermost workings of the brain. And indeed they have gone a long way towards that. Many have been discovered, including dopamine, noradrenaline, glutamate and 5-HT. More than 100 have been described.

In 1921, German pharmacologist Otto Loewi provided the first direct demonstration that neurons communicate via neurotransmitters. He mainly worked on, you guessed it . . . frogs, studying their vagus nerve, which controls digestion and heart rate. Loewi shares the credit for discovering acetylcholine (which Loewi called 'vagusstoff') with Henry Dale, sharing the Nobel Prize with him in 1936 – although given the major contributions frogs have made, perhaps Kermit should get a Nobel too?

Loewi often found inspiration for experiments while dreaming, including for the key experiment on the vagus nerve in frogs that identified acetylcholine. He dreamt of the experiment, woke up, scribbled it onto a piece of paper and fell back asleep. The next morning, he was devastated that he couldn't read his night-time scrawl. He said later that trying to remember the experiment was the longest day of his life. But fortunately, the following night he had the same dream, woke up, and went straight to the lab and did the experiment that would go on to win him the Nobel Prize. Some dream. Sadly that has never happened to me. My dreams are usually about me standing in front of a lecture theatre about to give the most important lecture of my life, except I'm naked, holding a monkey.

Loewi worked at the University of Graz but was forced out by the Nazis in 1938. He was jailed for three months but was released on the condition that he give up all his research and possessions. He moved to the US, where he stayed for the rest of his life.

Marthe Vogt was another German scientist who made important early contributions to the field of neurotransmitters. With the rise of the Nazis, she also moved, and went to work for Henry Dale. Dale gave her credit in his Nobel Prize speech for important work on acetylcholine that she had published in 1948. With the outbreak of war, she was sadly deemed a 'category A enemy alien' and was about to be interned when her colleagues in the UK came to her defence. She went on to do important work on two other neurotransmitters, serotonin and epinephrine.

The way neuroscientists described the brain began to make it seem like a fairground, with lights going on here and there, and neurons firing in many directions because of neurotransmitters, and action potentials running up and down like rollercoasters. Lots of electricity and noise and excitement. We now know that a single neuron can connect to up to 15,000 other neurons in a colossal network, and there are billions and billions of neurons. It's no wonder the mind is so complicated. A mere 867 million tweets are sent per day, according to recent figures, which is nothing compared to the communication going on in your brain every microsecond.

Different neurotransmitters were shown to do different things. Glutamate is what's called an excitatory neurotransmitter – it activates neurons. If it's overactive, it's been shown to cause damage, such as in stroke and epilepsy. It's also thought to be important for what happens to you when you have a hangover – apologies to any teetotal readers, but as an author I have to write about what I know. Alcohol has all kinds of effects on the body, but when it's in the brain it affects neurons, and thereby regulates neurotransmitters. For example, you relax when you drink alcohol because it represses glutamate, decreasing anxiety. The only thing is, when the alcohol is cleared from your brain, there is a rebound and you make higher levels than usual of glutamate. This makes you feel anxious, giving rise to so-called hangxiety.

But alcohol also *promotes* another neurotransmitter, GABA, whose job is to dampen brain activity, and this also leads to a decrease in anxiety. With alcohol you therefore get a double whammy – decreased glutamate and increased GABA, leading to an overwhelming urge to get on the dance floor, despite (at least in my case) an underwhelming level of coordination.

Dopamine is one of the more famous neurotransmitters because it makes us feel great. It regulates arousal and is linked to pleasure sensations; hence it is known as the motivation molecule. We evolved to yearn for new experiences and rewards, and dopamine reinforces particular activities when they have a successful or pleasurable outcome. This makes us want to keep doing those particular activities, like checking your iPhone. But there is a dark side to this. Drugs that bring us pleasure, like coffee or cocaine, or activities like sex, gambling or gaming all drive up dopamine, and lead to the 'high' that drives people to repeat the experience. And just like with alcohol, there is a rebound when you can't get a coffee or the PlayStation is broken, and levels of dopamine crash. This carrot-and-stick behaviour gives rise to addiction, as we crave more dopamine (carrot) and want to avoid the dopamine crash (stick). A further problem is tolerance, where we need more and more carrots for the same dopamine effect. All of this can wreak havoc in our brains. If dopamine levels are too low, or if we become tolerant to our own dopamine, this means we can get little pleasure from life, leading to depression and anxiety.

To kick an addiction, psychologists recommend abstinence from an addictive behaviour or substances, to allow dopamine levels to return to normal. They even suggest doing things that cause unpleasant feelings, like taking a cold shower, which will have a similar lowering effect on dopamine. One psychologist, Robert Sapolsky, got it right when he said, 'Well-timed deprivation can do wonders for pleasure.' Another effective way to restore dopamine levels to normal is to do something that allows you to achieve a 'flow state'. This is achieved by losing yourself in an activity such as music (my favourite), science (my other favourite), reading, exercise, meditation or even gaming, leading to a reduced focus on worry and self-reflection. The flow state seems to involve changes in neurotransmitters such as dopamine and norepinephrine in our brains, helping us because it means we aren't worrying as much. Activity of the dopaminergic pleasure system in our brains coincides with feelings of optimism, hope and reduced fatigue. Flow also means a relentless dedication to the task at hand. It's one reason why gamers can spend hours at a computer screen without feeling bored, tired or hungry. (Obviously, if the addiction you're trying to kick is gaming, more gaming is not going to have this positive effect.)

Serotonin (also known as 5-HT) is another neurotransmitter that is important for regulating mood. Low serotonin has been shown to be associated with depression and panic attacks. And just like with dopamine, achieving the flow state – be that through medication or otherwise – increases serotonin. Diseases like schizophrenia and depression are said to be caused by an imbalance in specific neurotransmitters, notably serotonin.

These insights not only help explain how our minds are controlled by neurotransmitters, but also how many drugs affect our brains. Cocaine prevents the uptake of dopamine back into neurons, which is why it makes people feel so good. Some antidepressants also target dopamine, while others like Prozac boost serotonin. The hallucinogen LSD overactivates the pathways that serotonin triggers in the brain, as well as dopamine. Quite how that translates into a psychedelic experience is not known – yet (but that's what's great about science). It might involve the opening of connections between different parts of the brain that don't normally connect. Nicotine, found in cigarettes, acts on certain forms of the acetylcholine receptor, which affects reward systems including dopamine levels in the brain, to promote wellbeing – obviously the tar in your lungs has a much less desirable effect on your body.

Brain diseases like Parkinson's and Alzheimer's have been shown to involve damage to specific parts of the brain, visible from post-mortems. In the case of Parkinson's disease, damage is in the substantia nigra, while for Alzheimer's it's the hippocampus. Just like the unfortunate Phineas Gage and his metal spike injury, this can tell us which part of the brain is important for movement (the substantia nigra) or memory (the hippocampus). In Parkinson's disease, the neurons that make dopamine in the substantia nigra are destroyed. One treatment is a chemical that can be turned into dopamine called L-Dopa, which increases dopamine.

Lovely Rita

Apart from work on neurotransmitters, there has been progress in our understanding of how the brain develops. This was the focus of the work of Rita Levi-Montalcini, an Italian neurobiologist. She worked in the anatomy department at the University of Turin but lost her job in 1938 when Jewish people in Italy were barred from holding university positions. She had a brief stint working in Belgium, but fearing the impending German invasion in 1939 she returned to Italy, to Florence, and set up a laboratory in her bedroom, to continue her work on the growth of nerve fibres in chicken embryos. When the Germans invaded Italy in 1943, she and her family were protected by non-Jewish families and avoided being deported to a concentration camp. She subsequently moved to the US, and at Washington University in St Louis she purified a substance from tumours that had lots of nerve tissue in them, and showed it could promote nerve growth, hence the name she gave it: nerve growth factor.

But before I forget, back to memory studies. The role of the hippocampus in memory formation was first observed in the 1950s, when a patient called Henry Molaison had part of his hippocampus removed in an effort to cure his severe epilepsy. He'd had a bicycle accident as a child and began having seizures. It cured his epilepsy but had the unfortunate consequence of preventing him from being able to form long-term memories. He was still able to complete crossword puzzles, but only from clues that needed information from before 1953 – prior to the removal of part of his hippocampus.

More recently, an important finding about this part of the brain was made by May-Britt Moser and her then-husband Edvard Moser. They identified neurons near the hippocampus in the entorhinal cortex – an area of the brain that allows us to know where we are in space. Damage to this part of the brain is one reason why patients with Alzheimer's disease fail to recognize where they are – they lose the memory of places that were previously familiar to them. Their so-called episodic memory is said to be affected.

We therefore know about the role of the hippocampus in memory, but the basis for memory is still not fully understood. From the 1960s, Eric Kandel worked on memory in the marine invertebrate *Aplysia californica* (California sea hare), which is a type of sea slug. Although primitive by our standards, they can still learn things. Eric was born in Vienna in 1929 and, like Otto Loewi, he and his family had to escape Nazi persecution and so left Austria for the US in 1939, when Eric was 10 years old. Later he went to Harvard and graduated with a degree in history and literature, but had taken classes in psychology too. He subsequently did a medical degree in New York but became very interested in neuroscience when he saw an oscilloscope being used to measure the action potential in neurons. He came across a method to isolate neurons from marine invertebrates, starting with the crayfish giant axon and moving on to the sea hare.

This would prove to be a genius move, as *Aplysia californica* has a mere 20,000 neurons (humans have an average of 86 billion), which can each be easily identified. The slug became the animal of choice for Kandel, as *Aplysia californica* can learn various things – most notably the gill-withdrawal reflex, which it uses to protect its fragile gill tissue from injury. Kandel was able to show that a molecule called cyclic AMP was made when the slug was learning, and the neurotransmitter serotonin increased cyclic AMP. The ultimate effect was to increase the production of proteins to alter synapses, making them stronger and laying down a memory. Kandel therefore got close to the molecular basis of memory. There was a change in the protein composition of synapses when learning was happening. If this process was blocked, *Aplysia* couldn't learn.

When he won the Nobel Prize for this discovery, the Austrians reclaimed him, calling it an 'Austrian' Nobel. This offended Kandel and the Austrian president called him to apologize. He asked what they could do to 'make things right'. Kandel said they should rename the street Doktor-Karl-Lueger-Ring, because Lueger was an anti-Semitic mayor of Vienna mentioned in Hitler's *Mein Kampf*. The president obliged and it was renamed as Universitaetsring (University Ring). Kandel then finally accepted an honorary citizenship of Vienna.

More recently, the concept of engrams has been proposed to explain memory, although it's hard to pin down exactly what an engram is (almost as hard as pinning down what NFTs and cryptocurrency are – more on those later), other than by giving the standard definition: a unit of cognitive information in a physical substance, yet to be fully defined. Although the hippocampus was shown to be important for memory, it turns out that memory is really something that is diffused over the whole brain. This was revealed by Karl Lashley, who progressively removed parts of the brains of rats but couldn't pin down exactly which part was central to memory, since removing any part made the rats forget in memory tests. Perhaps the hippocampus is the organizing centre of memory. It's thought that 'neural networks' or even single cells might make up an engram.

The Michael Jordan of Neuroscience

Today, a technique called optogenetics – first reported in 2005 by Karl Deisseroth at Stanford University – is being used to manipulate memories in mice and rats, and it might teach us how the mind actually works. Optogenetics involves introducing a gene encoding a protein into the brains of mice that will respond to a specific colour of laser light. Using special fibre-optic cables surgically implanted into the brain, shining a laser light on parts of the brain where the protein is expressed activates action potentials, which in turn triggers specific neuronal pathways in the brain depending on where you stimulate.

Deisseroth exemplifies what a scientist should be. He could see the amazing potential of this technology and so he spent 15 years developing it. He has said that early in his career he wanted to understand how feelings are constructed on a cellular level. He is a psychiatrist who treats patients with mental disorders like autism, schizophrenia and depression, and his overall goal is to help relieve the suffering caused by these conditions. He apparently rarely dresses in anything other than a T-shirt (untucked) and blue jeans. One of his mentors has called him 'the Michael Jordan or the Tom Brady or the Roger Federer of neuroscience'.

Before optogenetics, functional brain studies involved the

use of electrodes or drugs. Optogenetics allows neuroscientists to systematically manipulate specific neurons. Deisseroth first showed that, when activated, the light-sensitive proteins create an action potential in neurons in a Petri dish. He then went on to use this technique in living animals and set up workshops to teach other neuroscientists how to use it. Optogenetics has now become a standard tool of neuroscience, and researchers all over the world are using it to study learning, memory, perception, motivation, mood and appetite. All of these behaviours can be manipulated using optogenetics. It's been used to study the role of a part of the brain called the amygdala in the fear response, and to study cocaine addiction in the nucleus accumbens. It's also been used to create false memories in mice. Imagine if that were possible in humans! Neuroscientists are using optogenetics to study what goes wrong in diseases like Parkinson's disease, epilepsy, schizophrenia, autism, addiction, anxiety and depression. Deisseroth has most recently used optogenetics to identify a particular protein in the brain called HCN1, which when activated leads to what's called a dissociative state. This is a feeling that your mind is separate from your body. Your sense of self is somehow disrupted. It can happen during a traumatic event but can then recur as part of post-traumatic stress disorder. The protein implicated in the dissociative state could be a target for treatment. Deisseroth's experiments bring us a step closer to understanding how we have a sense of self at all – a sense of our own consciousness.

When asked to what he attributes his success, Deisseroth said 'curiosity'. He also said: 'We want to build a telescope to see a part of the sky nobody has ever seen.' Except he's not talking about astronomy, he's talking about the mind. His experiments, and those of other neuroscientists using optogenetics, bring us closer to knowing how neurons regulate memory and many other aspects of how our mind works, but there is a long way to go. Much work needs to be done to explain how memories are laid down or retrieved in the brain. One of the fastest supercomputers in the world had to use 82,000 processors to simulate human brain activity. And the brain is estimated to have almost as many neurons as there are stars in the Milky Way, with trillions of synapses. It took forty minutes of supercomputing to simulate one second of brain activity in a tiny patch of the brain, telling us what powerful things our brains are. It also tells us that we are nowhere near replicating the brain in a machine, which is just as well. If we did, like in *The Matrix* or *I, Robot*, those machines might try to take over the world.

And the big question remains: will we ever understand consciousness? Consciousness is an awareness of internal and external existence, and there are so many questions about it that are unanswered. Do animals have it? Is it mainly about our inner lives or also about how we think and perceive things? Do children have it, and if not, at what age do you get it? What happens when we are *un*conscious, either when asleep or in a coma? Many drugs profoundly affect consciousness, but how does that work? Then there is the 'stream of consciousness' as used by writers like James Joyce, who tried to capture what was going on in the minds of some of the characters in *Ulysses* as they walked around Dublin. Joyce used a technique in his writing where thoughts come and go, sometimes overlapping, triggered by what the character is seeing or remembering. Joyce had said that he was disappointed when he read Dickens because he felt the writing didn't convey consciousness (or even sub-consciousness), even though he recognized him as a great storyteller. In *Ulysses*, Joyce instead tried to convey what it was to be a thinking human being. Karl Deisseroth was a speaker at the 2018 conference commemorating the 'What is Life?' lectures by Erwin Schrödinger that I mentioned earlier in this book. He said to me that it is probably in Joyce's last book, *Finnegans Wake*, that we might find the basis for consciousness. I've tried reading it and it's completely impenetrable, so maybe Karl is right.

The famous German mathematician Gottfried Leibniz once said: 'If you could blow the brain up to the size of a mill and walk about inside, you would not find consciousness.' More recently, because of the uncertainties in the field, a neuroscientist friend of mine has said that discussing consciousness can damage your career. I'm not so sure. Big breakthroughs are coming.

9. Paranoid Android

I remember talking to my father about his childhood in the 1920s. He told me there were no televisions. I didn't believe him. My own children find it hard to believe that when I was a boy, computers were absent. I wonder what their children will have that they didn't have in their own childhood. It will more than likely involve artificial intelligence.

As a teenager, I was in the very first cohort of students in Ireland to study computers at school as part of the maths course. I was excited at being able to write a computer program in the BASIC language that could convert one currency into another. I remember the smell of the IBM Personal Computer 5150 that we used, the whirring of its cooling mechanism and the flashing cursor on the screen. A mistake in the program always gave rise to the phrase 'Syntax Error' coming up on the screen. When we formed our first punk rock band (we didn't get very far), guess what we called it?

The idea of artificial entities that can think goes back a long way. And of course, the mind has often been compared to a computer. In the sixteenth century, Paracelsus described how to make an 'artificial man'. This involved adding semen into horse dung and feeding it with human blood. Paracelsus said it would become a living baby. I'm not sure he ever tried it, but if he did, I'm pretty sure it didn't work and would have breached most health and safety regulations. There were some impressive inventions by then, such as the microscope, but also the pocket watch, the pencil and the flushable toilet.

But clearly, Paracelsus didn't have the technological foundations to build on that we have today. He was another scientist who became so famous that he could use one name – the Lizzo or Bono of his day. But his full name is worth recording: Philippus Aureolus Theophrastus Bombastus von Hohenheim. No wonder he went by Paracelsus.

In the thirteenth century, the Spanish philosopher Ramon Llull described devices that could use logic to make decisions. One involved two or more paper discs with alphabetical letters or symbols that linked to lists of attributes. The discs could be rotated to give rise to a large number of ideas, based on the lists referred to by the letters. Sort of like a particularly technical paper fortune-teller that children make today. As a young man, enjoying a rather bawdy life as a troubadour (nothing changes when it comes to musicians), Llull had a series of visions involving Jesus Christ on the cross suspended in mid-air. This led him to vow that he would write the best book in the world. I'll keep my motivations to myself.

In the late 1600s, German mathematician and philosopher Gottfried Leibniz figured that it would be easy to reduce argumentation to calculation, writing 'there would be no more need for disputation between two philosophers than between two accountants.' As a harbinger of what automation would bring hundreds of years later, he also said: 'It is unworthy of excellent men to lose hours like slaves in the labour of calculation which could safely be relegated to anyone else if machines were used.' Keep that in mind, all you excellent people, after a hard day sitting in front of a computer screen.

Leibniz's father was professor of moral philosophy at the University of Leipzig and Gottfried was allowed to read books in his father's library from the age of seven, which he did voraciously. Many of the books were in Latin, which he was fluent in by the age of 12. He entered the University of Leipzig at 14, graduated at 15 with a degree in philosophy, was awarded a master's at 17, and became a professor at 19, for work on the aforementioned Ramon Llull, so yet another example of how a previous scientist influences a later one. He also managed to obtain a law degree at the age of 19, but was turned down for a doctorate in law, which would have allowed him to practise as a lawyer, because he was too young.

His first paid job was secretary to the alchemical society in Nuremberg. But his main interest was mathematics, and in 1673 he made a calculating machine, which would add, subtract, multiply and divide. Known as the 'stepped reckoner', it had one section with 16 decimal digits and another with 8. It had a dial, like on an old-fashioned telephone, to set the multiplier digit, and a crank at the front to perform the calculation. The result from the calculation appeared in 16 windows at the back. Leibniz had got the idea from a pedometer, a device that counts steps, and an earlier and mechanically simpler calculator made by French mathematician Blaise Pascal.

Leibniz went to London and demonstrated the machine at the Royal Society in 1673. He fell out with Isaac Newton, who accused him of stealing his discovery of calculus – a branch of mathematics concerned with measuring continuous change – which Leibniz denied. This marked the beginning of a lifelong feud. Despite what their lofty egos might suggest, as we've seen time and again in this book: scientists are human. They bicker and fight and seek credit. On his way back from London to Germany, Leibniz stopped off in The Hague and met none other than the inventor of the microscope, Antonie van Leeuwenhoek.

He became friendly with female members of the nobility in Hanover, notably Sophia of Hanover and the Queen of Prussia. This was a good move, as he helped make the case for the House of Brunswick becoming the royal family of England, which duly happened, with George I becoming king followed by George II and George III. He started a history of the House of Brunswick going back to Charlemagne, but never finished it. Even though Leibniz had helped him become king, George I became so annoyed with him for not making progress that he forbade Leibniz from joining him in England. This might also have been because of the dispute with Newton.

Leibniz died in obscurity and was buried in an unmarked grave for 50 years until his work was rediscovered. Many in Europe felt that he had indeed plagiarized Newton when it came to calculus (he probably didn't) and that damaged his reputation, sadly overshadowing much of his other work. Einstein was a fan of Leibniz, even though he died some 200 years before Einstein's time, as he was the first proponent of the relativity of time and space, in direct contrast to Newton, who believed that time and space were fixed entities – another reason for Newton to dislike him.

It's Leibniz who set the ball rolling towards modern computers (literally, as some of the early computers had balls rattling around in their mechanisms), and his most important work in this regard concerned mathematical logic. He was one of the first to describe the binary numeral system. This came to him when he encountered the Chinese idea of yin and yang, which he said corresponded to 0 and 1. His description of a binary machine came very close to how modern computers work, at least 200 years before they were invented.

His rehabilitation was complete in 1986 with the establishment by the German government of the Leibniz Prize, an annual award of up to €2.5 million. An impressive total prize of €1001100010010110100000 – in binary numbering, that is. In 2018, Google celebrated his 372nd birthday with a Google Doodle of Leibniz's hand sketching 'Google' in binary ASCII code, the most common character-encoding format for text data in computers.

The Enchantress of Number

For the next advance we turn to the work of George Boole in thc nineteenth century. His work was directly inspired by Leibniz and would lay the foundations for the information age that we now live in. He was a self-taught mathematician who spent most of his working life at Queen's College, Cork (now University College Cork). Born in 1815, he was the son of a shoemaker in Lincoln, Lincolnshire. By the age of 16 he was the breadwinner for his parents and three younger siblings because his father's business had gone into decline. He became a teacher, working in Doncaster and Liverpool, and he established his own school in Lincoln at the age of 19. He took part in the campaign for the early closing of shops to reduce the number of hours worked in retail, and Sunday closing to give workers more time off for leisurely or learned pursuits.

Lincoln had a topographical society that Boole joined, and he began to publish on mathematical topics and started corresponding with mathematicians. In 1844, a mathematical paper he wrote entitled 'On a General Method in Analysis' won the first gold prize for mathematics at the Royal Society.

No mean feat for a teacher in Lincoln. Boole invented a type of algebra, subsequently named Boolean algebra, which at the time had no obvious uses but satisfied him greatly. It would later prove key to the development of digital computers, since it concerned mathematical operations on logical values with binary variables – meaning they are represented as 1 or 0, where 1 is 'true' and 0 is 'false'. In 1849, he was appointed professor of mathematics at Queen's College, Cork. Sadly, Boole didn't live to see the impact of his work. He died in 1864 from pneumonia, brought on when he walked three miles from his home to the university and gave a lecture in wet clothes. His wife looked after him but wrapped him in wet blankets. There was an old wives' tale that remedies should resemble their cause – not very logical, it must be said, and it certainly didn't work for Boole.

We are talking about Boole here, not Newton, but the apple did not fall far from the tree. His daughter Alicia Boole made important contributions to four-dimensional geometry. Another daughter, Lucy, became the first professor of chemistry in England, working at the Royal Free Hospital in London. A granddaughter, Joan Hinton, worked on the Manhattan Project that built the atom bomb in the US during the Second World War.

While the mathematicians were working away on logic, others were building machines. In the early 1800s, Charles Babbage designed what is now regarded as the first-ever digital programmable computer. He called it the difference engine.

Babbage had studied mathematics at Trinity College Cambridge, arriving self-taught in most aspects of contemporary mathematics. He found the teaching there disappointing. Inspired by Francis Bacon, an earlier student at Trinity – who, as we saw in the introduction, first proposed the scientific method in the early 1600s – Babbage formed a student society in mathematics called the Analytical Society. This was a remarkable group of students that included George Peacock, who would become an important mathematician working on the algebra of logic, and William Whewell, who would coin the term 'scientist'. The word 'science' had been used since at least the fourteenth century, and had the meaning 'the state of knowing'. Scientists were previously known as 'natural philosophers'. The first word Whewell came up with was 'savant', meaning a man of learning. But he decided against that because it was French, and that wouldn't do at all. He was inspired by the word 'artist' and so the battle between scientists and artists began. Which are you? I'm both. He did worry that 'scientist' would be linked to 'atheist', which was a fraught thing to be called at that time, but he finally decided on it anyway.

Shakespeare may have given us 'worthless', 'skim milk' and 'zany', but Whewell also gave us 'physicist', 'linguistics', 'electrode', 'ion', 'anode' and 'cathode'. Some list. I came up with one word, Mal, that stands for MyD88-adapter-like, which was discovered in my lab in the early 2000s. Mal is a molecular switch in macrophages, the front-line cells of your immune system, which is turned on in response to infection, triggering the macrophage to fight. You are now one of probably a couple of hundred people who know the word I invented.

The Analytical Society used to meet regularly for breakfast, discussing findings in mathematics and science. It was a Bacon-inspired society, but sadly it's not known if they enjoyed bacon at their breakfasts.

Babbage was also an avid member of the Ghost Club, which investigated supernatural phenomena. It was founded in London in 1862 and included in its membership Charles Dickens and Sir Arthur Conan Doyle, of Sherlock Holmes fame. The Ghost Club would organize seances and attempt to prove the existence of ghosts, which sounds more fun than the potentially bacon-less breakfasts of the Analytical Society. He was obviously a fan of clubs because he was also a member of the Extractors Club, whose only goal was to liberate any of its members should they be committed to an asylum. Quite why they were so worried about that isn't known. Might it have been that scientists were often seen as crazy?

After Cambridge, despite much effort applying to several universities, Babbage failed to secure a professorship, probably because he was an irascible character. He lived off an allowance from his rich father, upon the death of whom he inherited the equivalent of £9 million today. He was finally appointed to the Lucasian Professorship of Mathematics at Cambridge in 1828, previously held by Newton, having failed in his attempt at the post three times previously. You've got to hand it to him for persevering. But he was heavily criticized by colleagues for not giving any lectures. He was prevented from giving a lecture on how Cambridge should be reformed to allow staff to carry out more research and do less teaching, and also to apply that research in useful ways. But that is exactly what happened in many universities in the latter half of the twentieth century, more evidence that Babbage was well ahead of his time.

Babbage seemed to be always complaining, notably about the state of science in England, and he tried to oust Davies Gilbert from his position as president of the Royal Society. He wrote a lot about 'nuisances' and had a particular hatred of street musicians, notably organ grinders – people who played a barrel organ in the street, turning a crank to operate the bellows – writing: 'It is difficult to estimate the misery inflicted upon thousands of persons . . . by organ-grinders.' He was criticized in the House of Commons in 1864 for 'commencing a crusade against the popular game of trundling hoops'. This was a popular pastime for children and involved rolling a large hoop along the ground using a stick. He regularly reported his neighbours to the police, and so they had their children harass him in the streets, left dead cats on his doorstep, and they even put together a brass band to play outside his house at 1 a.m.

His reputation as a misery guts seems well founded, since he was more interested in analysing jokes than enjoying them, writing 'it has often struck me that an analysis of the causes of wit would be a very interesting subject of inquiry. With that view I collected many jest-books, but fortunately in this one instance I had resolution to abstain from distracting my attention from more important inquiries.'

And yet Babbage is most interesting to us because he came up with the difference engine. He had been motivated to invent an accurate way to carry out astronomical and mathematical calculations, by coming up with a mechanical calculating machine. In 1822, he made a small difference engine, which used the decimal system and was operated by a crank handle. The British government became interested and gave him a grant of £17,000. The challenge at that time was that it was not economical to make the metal parts with the precision and in the quantity needed.

Lady Byron, wife of poet Lord Byron, had a strong interest in mathematics, and in 1833 she went to see prototypes. 'We went to see the "thinking machine" (or so it seems) last Monday. It raised several numbers to the 2nd and 3rd powers and extracted the root of a quadratic equation.' Lady Byron had been a gifted child, and her parents had hired a former Cambridge professor, William Frend, as her tutor. She was given an unusual education for a woman at that time, which covered classical literature, philosophy, science and mathematics – her favourite subject. This interest led her husband to give her the charming nickname 'Princess of Parallelograms'.

Babbage's difference engines were never successfully built in his lifetime, but like today's computers involved information processing. To celebrate the 200th anniversary of his birth in 1991, the Science Museum in London had one made from Babbage's original drawings. It worked! But it weighed five tons and was seven feet high – one possible reason why no one is wearing a Babbage watch.

But then the story of computers gets even more interesting. Lady Byron's daughter was called Ada Lovelace and she was brilliant. She took what Babbage had proposed and developed it further towards the computers we know today. She goes down in history as the first computer programmer.

Lovelace was the only child of Lady Byron and Lord Byron. Her mother was fearful that she might turn out like her father, who rebelled against social norms, including a rumoured incestuous relationship with his half-sister while married. Her mother therefore decided to give Lovelace an education mainly in science and mathematics, to discourage her from studying literature, which she felt might lead her astray.

When Lovelace was five weeks old her parents separated, and she never saw her father again. She was mainly brought up by her maternal grandmother, as apart from organizing her education, her mother had little input in her life. In one letter Lady Byron refers to her daughter as 'it': 'I talk to it for your satisfaction, not my own, and shall be very glad when you have it under your own.' Nice.

As a child, Lovelace was often ill and almost died of measles. She spent a whole year in bed convalescing, and rather than idling it was in that time that her interest in mathematics grew. When Lovelace was a teenager, Lady Byron had her watched by friends who reported on any immoral behaviour, although quite what that would be wasn't described. Lovelace called them the 'Furies' and said they invented stories about her.

At the age of 18 she had a relationship with her tutor and tried to elope with him, but his family intervened. In one act of good parenting (and in fairness, Lord Byron was entirely absent) Lady Byron organized a new tutor, Mary Somerville. This was a stroke of genius, because Somerville was one of the few female scientists in England at that time, and she proved to be a huge influence on Lovelace. Somerville was a remarkable woman. She had studied mathematics and Latin as a child, despite much resistance to the teaching of those subjects to a girl, and she refused to take sugar in her tea as a protest against the slave trade. In 1868, hers was the first signature on the first petition to Parliament to give women the vote. Along with Caroline Herschel, she was one of the first female members of the Royal Astronomical Society, and in 1834 she was the first woman to be called 'scientist' in print.

Somerville had a wide circle of scientific friends, including Charles Babbage. She took Lovelace to a gathering where Babbage spoke about his difference engine and Lovelace was transfixed. Lovelace also met Augustus De Morgan, a famous mathematician at that time, and he proved supportive, writing to Lady Byron that Lovelace might become 'an original mathematical investigator, perhaps of first-rate eminence'. Lovelace began working on mathematical problems and would often walk to Somerville's house nearby to discuss her work.

Lovelace began to visit Babbage regularly to watch his progress on his difference engine. He was impressed by her abilities and called her the 'Enchantress of Number'. There seems to be a theme of patronizing nicknames for the brilliant women of that family, unless you take the Princess of Parallelograms and the Enchantress of Number as some sort of early badass scientific superheroes. On her visits to see Babbage, they would sometimes go for a walk, and the terrace they walked along has been renamed the Philosopher's Walk because of the many mathematical discussions they had there.

An Italian mathematician, Luigi Menabrea, who would go on to be prime minister of Italy, had written an article on Babbage's follow-up machine to the difference engine, which was called the analytical engine. Lovelace translated the article with Babbage's help, and included a set of notes that outlined her algorithm. A copy was sold in 2018 for £95,000. The notes were around three times longer than the article itself. Note G contains a step-by-step description for the computation of Bernoulli numbers, a complex problem to rapidly find the sums of the powers of different numbers first outlined in a series of tables by Bernoulli. He said, 'with the help of this table, it took me less than half of a quarter of an hour to find that the tenth powers of the first 1,000 numbers being added together will yield the sum 91,409,924,241,424,243,424,241,924,242,500.' Lovelace's algorithm to solve Bernoulli numbers is recognized as the first computer program, although Babbage had written what could be considered algorithms – defined as a set of rules to be followed in mathematical calculations – for the analytical engine, prior to the notes.

But Lovelace realized that the analytical engine could extend well beyond number crunching, which was the main thing that Babbage was concerned with. She wrote that the analytical engine 'might act upon other things besides *number*, were objects found whose mutual fundamental relations could be expressed by those of the abstract science of operations . . . Supposing, for instance, that the fundamental relations of pitched sounds in the science of harmony and of musical composition were susceptible of such expression and adaptations, the engine might compose elaborate pieces of music.'

In her thirties, Lovelace developed an interest in gambling, which at the time was seen as scandalous for a woman. In 1851, she formed a syndicate with male friends and came up with what she thought might be a foolproof mathematical model to ensure wins using large bets. It didn't work and she had trouble clearing her horse-racing debts.

When she was dying of cancer at the early age of 36 in 1852, she wrote to Babbage asking him to be her executor. Since 2015, drawings of Babbage and Lovelace are on all British passports and there are many awards named after her. Ada Lovelace Day falls on the second Tuesday in October and has the goal of raising the profile of women in science, technology, engineering and mathematics.

Finding the Wolfpacks

We have to wait until 1909 for the next step in our journey towards computers. Percy Ludgate was an accountant in Dublin. He was working as a clerk at a corn merchant, but at night he was designing analytical engines, and doesn't seem to have been aware of Babbage's work. His designs were similar to those of Babbage, although he introduced some new elements, including what is seen as the first program-control device (now called a subroutine): a sequence of programming instructions to perform a specific task. Not much else is known about him, and he died at the age of 39.

By the 1930s, electrical engineers had devised ways to use electrical switches to do logic, where Babbage and Ludgate had used mechanical methods. Credit for the first attempt goes to Spanish civil engineer Leonardo Torres y Quevedo. Between 1914 and 1920, he demonstrated that all the cog-wheels of Babbage's analytical engine could be replaced with electromechanical parts. His 1920 machine, the electromechanical arithmometer (admittedly not a catchy name), could perform arithmetic operations represented in decimals, using a typewriter to send commands and printing the results. Sounds like a computer, right – although it would take another 60 years for the first laptop to be invented. Torres y Quevedo was a pioneer in two other areas. One of them would literally take off: radio control. He first demonstrated it in 1906 in the port of Bilbao, guiding a boat remotely from the shore with people aboard. The other area was what we now call gaming. He invented the first computer game – an automaton that could play chess against a human opponent. Electrical and then electronic switches would go on to form the basis for all electronic digital computers.

The next important advance happened in the 1940s, with computers moving away from performing complex calculations based on logic, and instead being deployed in codebreaking, with the development of machines such as Z3, ENIAC and Colossus. British mathematician Alan Turing played a pivotal role in these developments, and computer science in general. He was born in Maida Vale, London, in 1912. The head of the primary school he attended noted that she had 'clever boys and hardworking boys, but Alan is a genius'. He went to secondary school in Sherborne in Dorset, and his first day coincided with the 1926 general strike. He was so keen to attend that he rode his bicycle the 60 miles from Southampton to get there. At 15 he was able to solve advanced mathematical problems, and at 16 he read Einstein's work, grasping it all immediately – and, let's be honest, lots of people *never* grasp it. He obtained a first-class honours degree in mathematics from King's College Cambridge, becoming a Fellow upon graduation. At the age of 24 he wrote what has been called 'easily the most influential math paper in history' in which he described a 'universal computing machine' which could perform any mathematical computation using algorithms. This became known as the Turing machine. The paper is acknowledged as the first to describe the modern computer as we know it.

Turing then spent time at Princeton and worked on cryptology – the science of codes. He returned to Cambridge and attended lectures by famous philosopher Ludwig Wittgenstein about the foundations of mathematics. Turing interrupted constantly, and the lectures were apparently so compelling and detailed that they were later reconstructed verbatim by students who had been there, including Turing's interruptions.

In 1938, Turing worked with the UK's Government Code and Cypher School, who were tasked with cracking German codes. The Germans had a machine for generating secret messages called Enigma, and Turing tried to understand how it worked. On 4 September 1939, the day after the UK declared war on Germany, Turing joined the staff at

Bletchley Park, Britain's codebreaking centre. He was viewed as a mathematical genius and something of an eccentric, and was known to his colleagues as 'Prof'. In the summer he would cycle with a gas mask on to ward off hay fever. His bicycle had a fault, with the chain coming off regularly, but instead of having it fixed he would count the number of times the pedals went around, and then get off in time to adjust the chain by hand. He was a marathon runner who would sometimes run the 40 miles to London for meetings. He said of his running: 'I have such a stressful job that the only way I can get it out of my mind is by running hard; it's the only way I can get some release.' He was such a competitive runner that he tried out for the 1948 British Olympic running team but was held back by injury.

At Bletchley Park, he made great progress. The so-called wolfpacks – groups of German U-boats – were causing havoc at sea, sinking Allied and civilian ships left, right and centre. But by the summer of 1941, codebreaking that revealed the positions of the U-boats meant that shipping losses had fallen from 282,000 tons a month to less than 64,000 tons. But Turing wanted to do even better. He was frustrated by the lack of staff and resources. He wrote directly to Winston Churchill, stating that their need was small compared to the vast sums being spent on men and equipment. Churchill responded immediately, writing to his chief assistant: 'ACTION THIS DAY. Make sure they have all they want an extreme priority and report to me that this has been done.' From then on, Bletchley Park was well supported. Turing brought mathematician Tommy Flowers on board, and Flowers would go on to design the world's first digital electronic computer, Colossus.

Turing also worked with Joan Clarke, an outstanding

mathematician who came to Bletchley with a double first in mathematics from Cambridge. But she was denied a full degree because women were barred from this until 1948. One of her tutors at Cambridge, Gordon Welchman, had noticed her remarkable abilities and recruited her to join him at Bletchley Park. When she got there, she was initially placed in an all-women group called The Girls, who mainly did clerical work. She became a key member of Turing's team. Groups of activities were isolated in individual huts at Bletchley, to maintain secrecy. For her work she was promoted but only to deputy head of the hut she was in, because she was a woman. She was also paid less than the men, which was standard at that time – and which persists in a lesser form today. There is still work to be done in that regard.

She and Turing became good friends. They shared many hobbies and friends reported that they had similar personalities. Turing arranged shifts so that they could work together and, in 1941, Turing proposed marriage. He admitted his homosexuality but she said it didn't faze her.

While at Bletchley Park, Turing became worried that should there be a German invasion he might lose his savings. He bought two silver bars worth £250 (they would be worth around £56,000 today) and buried them in the Bletchley Park forest. When the war ended, he went to dig them up but he was unable to break his own code describing where he'd buried them, which must have been frustrating for the arch-codebreaker himself. They remain buried.

Turing's work and that of his colleagues at Bletchley Park are estimated to have shortened the war by two years and saved over 14 million lives. A great example of brain over brawn, and the usefulness of mathematics.

After the war, Turing moved to London and worked at the

National Physical Laboratory on the first computer that could store program instructions in its memory. He then moved to the mathematics department at the Victoria University of Manchester. While there, he wrote theoretical papers on artificial intelligence, and came up with the Turing test. This was an attempt to determine how to assess if a machine was 'intelligent'. The idea was that a computer could be deemed intelligent if a human interrogator couldn't tell if its answers to questions were coming from a human or a computer. Those annoying CAPTCHA tests on the internet aim to do just that – determine if you are a robot or a human, although how it manages to do that is well beyond me. Surely a robot can spot tractors or zebra crossings?

Following Turing's conviction for homosexuality – a crime in the UK at the time – he was subjected to cruel hormonal treatment to avoid jail time, which likely led to depression. Turing died by suicide at the age of 41. He was found by his housekeeper, who saw a half-eaten apple by his bed – he had died of cyanide poisoning, which was possibly injected into the apple. It has been speculated that he was re-enacting a scene from his favourite fairy tale, *Snow White and the Seven Dwarfs*. He was known to have particularly enjoyed the scene where the wicked queen immerses her apple in a poisonous brew.

However, there are others that think his death was accidental. Turing had an apparatus in his room for electroplating gold onto spoons. It used potassium cyanide, and it's possible he inhaled some accidentally. His biographer Andrew Hodges has speculated that Turing ordered the equipment so that his mother wouldn't have to think he had committed suicide but instead was careless with laboratory chemicals. There is also a rumour that the Apple logo on your iPhone or Mac is a tribute to Turing, with the bite taken out of it representing the bite Turing took from the poisoned apple. Whatever the truth of the matter, Turing didn't live to see the enormous contribution he made to the development of computers – and his barbaric treatment at the hands of the British justice system is a nasty part of this story.

In 1999, *Time* magazine named Turing as one of the 100 most important people of the twentieth century: 'The fact remains that everyone who taps a keyboard, opens a spreadsheet or a word-processing program is working on an incarnation of a Turing machine.' A petition signed by 37,000 people asked for a pardon for Turing, which was granted in 2012. What became known as the Alan Turing law led to the pardoning of 75,000 others in the UK who had been convicted for acts of homosexuality.

Flame-Throwing Trumpets

While Turing was busy inventing computer design, Claude Shannon in the US was working on information theory – the mathematical study of the coding of information using symbols – and it would prove critical for the development of computers. Shannon was born in 1916 in Michigan, and as a child developed an interest in electrical devices. He made a radio-controlled model boat and a telegraph to communicate with a friend who lived half a mile away. His hero was the American inventor and businessman Thomas Edison, who he discovered later to be a distant cousin.

Shannon attended the University of Michigan and there he came across the work of George Boole. He graduated with two degrees, one in electrical engineering and the other in mathematics, before moving to the Massachusetts Institute of Technology for graduate work – the leading institution in the US for technology. For his master's thesis, he designed a computer using switching circuits based on Boolean algebra. This work is the basis for all electronic digital computers. Noted Harvard psychologist Howard Gardner has called it 'possibly the most important, and also the most noted, master's thesis of the century'. In it, Shannon described how the binary digits (0 and 1) are the fundamental element in all communication and information processing. Boolean algebra states that a statement of logic carries a 1 if true, and a 0 if false; Shannon proposed that a switch in the 'on' position would equate to a Boolean 1, while in the 'off' position it was 0. By reducing all information to a series of ones and zeros, Shannon realized that information could be processed using on-off switches.

After obtaining his PhD he moved on to the Institute for Advanced Study at Princeton, where Einstein worked, and then on to Bell Labs, where in 1943 he met Turing. They would meet up regularly in the canteen at teatime, and Turing showed him his paper on the Turing machine, which had a big influence on him. Regular tea breaks are vital for science.

Like Turing, Shannon worked on cryptology and code-breaking, and that work greatly informed his subsequent work on communication theory – he said they were 'so close together you couldn't separate them'. Shannon had numerous hobbies, including juggling, unicycling and chess. He invented a juggling machine, a flame-throwing trumpet (for reasons still unclear), and a device that could solve any Rubik's cube. He is remembered in a cryptocurrency named after him, the shannon. He once wrote: 'I can visualize a time in the future when we will be to robots as dogs are to humans . . . I am rooting for the machines!' Something to look forward to.

In 1956 Shannon attended a conference known as the Dartmouth Workshop. It was organized by a young academic, John McCarthy, who was working at Dartmouth College. Nine scientists and mathematicians attended. The workshop came up with the term 'artificial intelligence' and made the following assertion: 'Every aspect of learning or any other feature of intelligence can be so precisely described that a machine can be made to simulate it.' The race was on to make such a machine. In 1965, Gordon Moore came up with his law – the so-called Moore's law – that stated that there would be a doubling every two years in what effectively is the memory that a computer can hold. This has largely held true.

Steve Jobs Gets Stoned

Advances in computer technology proceeded apace from the 1960s onwards, with the advent of personal computers being a key development, as first configured by Steve Wozniak and Steve Jobs.

Wozniak was born in 1950 in San Jose, California. He was a *Star Trek* fan in his youth, and credits this with inspiring him to work on computer design. He first went to college at the University of Colorado but was expelled in his first year for hacking into the university's computer system. He transferred to the University of California, Berkeley in 1971 to study engineering, and in his first year built his first computer. He named it Cream Soda after his favourite beverage. It seems he then got bored, dropped out of Berkeley and got a job with Hewlett-Packard designing calculators. Steve Jobs was working there too, and they struck up a friendship.

Wozniak read an article in *Esquire* magazine entitled 'Secrets of the Little Blue Box'. Blue boxes were devices that mimicked the sound tones that controlled access through the telephone networks, and allowed you to make long-distance phone calls at no cost. Wozniak designed one, and Jobs took on the task of selling them to so-called phone phreaks – which was illegal. They managed to sell more than 200 at $150 each, and split the profits. Jobs said that if it weren't for Wozniak's blue boxes, they would not have gone on to found Apple.

Steve Jobs was born in San Francisco in 1955 to a Syrian father and American mother, but was given up for adoption. His adoptive mother, Clara Jobs, said that for the first six years of Jobs's life she was too frightened to love him in case she had to give him up. Not that Jobs noticed. He said he felt deeply loved by his adoptive parents. But he was disruptive at school and was always carrying out pranks on his teachers. As a teenager he attended Homestead High School, which had strong links to Silicon Valley – a region that had already established itself as the global centre for technology and innovation.

Jobs said that in the mid-1970s he 'got stoned for the first time' and this changed him. He developed a strong interest in literature, particularly Shakespeare, but also electronics. He attended Reed College in Portland, Oregon, but dropped out after one semester, going back to his parents' house. He got a job at Atari, Inc., where he was assigned the task of creating a circuit board for the arcade video game *Breakout*. Atari offered him $100 for each chip that was eliminated in the machine. Jobs asked Wozniak to help, and to the astonishment of the engineers at Atari, Wozniak reduced the chip count by 50 – which was thought unachievable. But Jobs didn't tell Wozniak about the bonus and never paid him for his efforts.

In 1975, Wozniak began designing the Apple I computer. On 29 June they tested the first prototype, displaying a few letters on the TV screen attached – the first time in history that characters appeared on a screen from a home computer. They began working in Jobs's bedroom at his parents' house, subsequently moving to the garage. Jobs came up with the name Apple after spending time at a farm commune in Oregon which had an orchard – although the bite out of the apple might have been inspired by Turing. They initially planned to make circuit boards to sell to computer enthusiasts, but instead constructed the first Apple computer, which consisted of a circuit board, TV set, Panasonic cassette-tape recorder and a keyboard. They tried on five separate occasions to sell the design to Hewlett-Packard but were turned down; apparently HP didn't see why an ordinary person would need to use a computer. When Wozniak presented it to them, they saw it as something only a computer enthusiast (or 'nerd') would want to use.

They founded the Apple Computer Company (now called Apple Inc.) and began to raise money. Funding for their first circuit boards came from the sale of Wozniak's scientific calculator and Jobs's Volkswagen van. One investor named Arthur Rock gave them $57,000 when he saw the crowded Apple booth at the Home Brew Computer Show, an event for computer hobbyists. The Apple I retailed at $666.66. Wozniak later said he'd had no idea that this was the mark of the beast from the Bible, but instead came up with the price because he liked repeating digits.

By April 1977, Jobs and Wozniak introduced Apple II at the West Coast Computer Faire, and it went on to be the first mass-produced microcomputer in the world. By 1980, 25-year-old Jobs and 30-year-old Wozniak were millionaires, and Arthur Rock's shares were worth $14 million when the company went public. At Apple, Wozniak was the innovator and Jobs was more on the side of business development and marketing. Wozniak has said: 'Everything I did at Apple that was an A+ job and that got us places, I had two things in my favour . . . I had no money and I had had no training'. Wozniak did eventually go back to college, completing his degree at UC Berkeley in electrical engineering and computer science under the pseudonym Rocky Raccoon Clark.

We eventually got to computers small enough to fit in the palm of your hand that were as powerful as any computer built before the 1980s. However, to become all-pervasive we needed something else. Something global and something that connected up all of the computers in the world, in some sort of web . . .

Ta-Da! The Internet

British scientist Tim Berners-Lee was born in London in 1955 to parents who were computer scientists. His main hobby as a child was trainspotting and he first became interested in electronics by working on a model railway. While working in Switzerland at CERN (the European Organization for Nuclear Research), he came up with the World Wide Web, an information system which allowed documents and other resources to be shared over the internet. Larry Page and Sergey Brin would start out exploring link structures in the World Wide Web, which in turn gave rise to Google. When asked how he did it, Berners-Lee said: 'I just had to take the hypertext idea and connect it to the TCP and DNS ideas and – ta-da! – the World Wide Web!' Ta-da indeed. What are you talking about, Tim?

The internet works by computers connecting to each other via wires, cables, radio waves and other types of infrastructure. The data is translated into packets of light or electricity (bits) and then interpreted by the computer that receives them. The bits are moved at the speed of light, and the more bits that can be passed, the faster the internet works. A good comparison is the way the Statue of Liberty was constructed in the US. It was built in France but was too big to transport and so was shipped to the US in pieces, with instructions as to how to assemble it. It took a long time to assemble. A photo of the Statue of Liberty as a series of bits in packets can be sent over the internet and assembled in milliseconds.

Berners-Lee published the first website, dedicated to the World Wide Web itself, on 20 December 1990. However, there were precursors to the World Wide Web, stemming from the work of Karen Catlin and Janet Walker. Robert Cailliau organized the first conference on the Web in 1994. Search engines like Google were invented, which revolutionized how we find information and then spread it. Most, if not all, of the knowledge that humans have accumulated through history (and everything in this book) can now be found in an instant in your pocket. It's simply astounding. So it is no surprise that in a global poll commissioned by the British Council of 25 eminent scientists, academics, writers and world leaders, the invention of the World Wide Web has been listed as number one in the top 80 cultural moments that shaped the world. Number two was the discovery of a method to mass-produce penicillin, and number three was home computers.

In 1998, the first search using a search engine was the name 'Gerhard Casper', the then president of Stanford University, where the inventors of Google – Larry Page and Sergey Brin – were students. I wonder what they found way back then? In 2022, the most searched terms in order were Facebook, YouTube, Amazon, weather and Walmart. Google was founded in 1998 and, given the sci-fi film, I carried out my first search in the highly appropriate year 2001. Afterwards, I walked past the Old Library at Trinity College Dublin. It opened in 1732 and at that time was reputedly the biggest library in the world. Most human knowledge that was in books could have been found in that library. When I was a student in the 1980s, you went there, found a book and searched the index. And now in 2001, on a computer, a simple search with keywords revealed the knowledge I was looking for. I thought, how magical is that, and what would the Irish monks of the eighth century, who wrote out the Book of Kells, then seen as the repository of much human knowledge, make of that? Clearly the work of Satan.

By 1997, the Deep Blue chess computer became the first that could beat the best human players, famously beating reigning world champion Garry Kasparov. Social media began to dominate our lives in the mid-noughties, for good and for ill, with billions of people using Facebook, Instagram and Twitter to communicate and obtain information. Two billion people use Instagram; Facebook has almost 3 billion users and Twitter 450 million. No part of our lives has been left untouched by these developments, with computer technology dominating our personal lives, taking over the world of entertainment and even that most human of activities, dating. Warfare is increasingly dominated by drones, computer-controlled weapons and cyber-warfare, with less need for humans in the armed forces. Though the recent war in Ukraine tells us two things: 1) war done by intelligent machines isn't there yet and 2) human intelligence still hasn't worked out that war is stupid.

We live in a digital world. We entertain ourselves digitally – which works in both senses: computers are everywhere and we use our fingers to operate them – via Netflix and Spotify. We fight digital battles on Fortnite. We meet digitally on Zoom – that all-pervasive way of interacting with people that took off during the COVID-19 pandemic. We work from home with multiple digital programs. Even when we exercise, our digital watches tell us when to push a bit harder. The truth of the matter is, almost everything we do is done digitally. The life I lead now would have looked like something out of a science fiction movie to me as a child.

Fungible and Non-Fungible

Digital developments were revolutionizing the way we lived and it was only a matter of time before things like ownership of something – or the right to use something – would become digital too, and that's what's happened with the non-fungible token – or NFT for short. NFTs are digital assets that are completely unique. They can be used to represent things such as photos, videos, audio or other types of digital files.

The term 'fungible' is used when it comes to assigning value to a digital entity. Something is fungible if one unit of it is equivalent to another unit. Money is fungible: one

10-dollar bill can be exchanged for another 10-dollar bill. 'Non-fungible' means not mutually exchangeable. You might hold 10 NFTs linked to something, but they won't be the equivalent of 10 NFTs linked to something else. They are therefore a bit like different currencies, where one day $1 might buy you £1 and the next day £1.50.

Cryptography, as worked on by the likes of Alan Turing and Joan Clarke at Bletchley Park, is used as the way to verify the value of the NFT. Unique information is stored in a blockchain, which is a list of recorded numbers that are linked together by cryptography. Each block has a record of the previous block and a timestamp. It's called a blockchain because of the links between the blocks. Each block reinforces the one before, and so are resistant to modification. I'll bet Turing and Clarke never imagined that their work would lead to NFTs.

NFT expert Alex Salnikov has said that the 'digital signature is one of the greatest inventions of the twentieth century'. All NFTs are attached to your own private identifier, providing the evidence that you own them. NFTs are like private property but they only exist digitally. In 2020 the NFT market reached $100 million. In 2021 it hit $41 billion. An example Salnikov gives is, say you want to buy a new Star Wars Lego toy. You're not buying it like you would say a kitchen knife, which has a practical use. You're buying it because you like it and want to play with it. There's no practical reason for it to exist in the physical world, if instead you can play with it virtually. It's the same if it's an NFT – it's yours to play with in the augmented reality (AR) world, where experience is enhanced by computer-generated information.

The number of people using NFTs is growing exponentially. A digital work of art called *Everydays: the First 5000 Days*, by artist Mike Winkelmann, was sold for $69.3 million plus a $9 million fee for the auctioneer. NFTs can be used by musicians to sell and profit from their music. Kings of Leon were the first band to release a new album in the form of an NFT that carried access to artwork, music and a chance to win prizes. It generated $2 million in sales before the unsold NFTs were 'destroyed'. The band donated $500,000 of the proceeds to support music crews during the pandemic.

In May 2021, UC Berkeley announced that it would be auctioning NFTs for the patents covering the CRISPR-Cas9 gene-editing technology. This is a collectors' item whose provenance is assured from its NFT. They were sold on 8 June 2021 for just under $55,000. An NFT of the source code of the World Wide Web written by Tim Berners-Lee was sold for $5.4 million. One wonders what it would have sold for if it had been written by hand by Tim himself. Perhaps it's an age thing, where younger people who grew up with the internet are happy to own things that others can have too.

Our lives will continue to be changed by artificial intelligence and computer technologies in ways we can't imagine. Unfortunately, this isn't without problems. A major and malign problem is how social media can be used to manipulate people, spread misinformation, and promote anti-scientific rhetoric. This was perhaps anticipated by George Orwell in his book *Nineteen Eighty-Four*, which was published in 1949 and in which he predicted the use of Newspeak – disinformation spread through telescreens to control the population. It was also discussed by Bill Gates in an interview with science fiction author Terry Pratchett in 1995, although Gates thought any disinformation would be rapidly corrected. Not quickly enough, Bill . . . But then again, it's not as if there hasn't always been propaganda. Before the internet, it was the radio, or the leaflet.

A recent survey has revealed that over half of all scientists who did media work during the COVID-19 pandemic were trolled and attacked on social media, and that includes me. This is dangerous. All of the amazing advances we have read about in this book came about because scientists were able to do science. That's not to say they were never held back – Galileo had it worse than a bit of online trolling – but not since the Renaissance have scientists been attacked on such a large scale. I am a big advocate for scientific literacy – it allows people to evaluate what they are reading in the digital world, and make their own minds up from a position of knowledge and reason. Science is less scary if you understand it.

That most eminent of physicists Stephen Hawking – he was even in *The Simpsons* – said that the development of full artificial intelligence could 'spell the end of the human race'. Now there's a cheery thought. He speculated that AI could take off on its own and redesign itself at an ever-increasing rate. However, I am optimistic that we can use AI to solve many of the world's problems by analysing and interpreting the massive data sets needed, for example, to understand climate change and its cause, or to improve drugs for major diseases. Otherwise, what will all this effort – from Leibniz to Boole to Babbage to Somerville to Lovelace to Turing to Shannon to Wozniak to Berners-Lee and many people in between – have been for? As ever, it is in our hands. Like Shannon, I will be tooting my flame-throwing trumpet while singing, 'I'm rooting for the machines' – but only if they are working to improve our lives and the health of our planet.

10. A Hard Rain's Gonna Fall

An unintended consequence of our technological evolution and the booming population it has supported has been an exponential need for energy and resources to build our cities, provide our food, and fuel our cars and planes. Thomas Newcomen and James Watt might be hugely lauded for their work on the steam engine, but that engine and the ones that followed have put pressure on the planet because of carbon emissions.

Before the Industrial Revolution, carbon dioxide levels were around 280 parts per million (ppm, meaning mg/litre of air) for almost 6,000 years. Since then, we humans have generated around 1.5 trillion tons of carbon dioxide pollution. The levels are now estimated to be 400 ppm, which is the same as it was 4.5 million years ago – a recent high point for carbon dioxide in the air. Recent because the Earth is 4.5 billion years old. Back then, temperatures were 7 degrees Fahrenheit higher, sea levels were 5–25 metres higher (which would mean most of the world's biggest cities would now be underwater), and there was a large forest in the Arctic tundra. We are therefore heading back in that direction. The combined impact of 8 billion people using energy mainly derived from burning fossil fuels, and extending our footprint to cover more and more of the Earth's land and sea, has become a problem that we cannot ignore.

The main problem is the burning of fossil fuels, which is responsible for over 65 per cent of greenhouse gas emissions, but also agriculture, which contributes over 20 per cent, much of this in the form of methane from belching cattle. I am convinced that we can do something about this problem, but we will need to use our collective brilliance to do it. We need to use all our votes and our powers of persuasion to convince governments and everyone to change their use of fossil fuels and agricultural practices. Our ingenuity caused this problem, and we need to use our ingenuity to crack it. We've always wondered how the world as we know it might end, and it's certainly been the subject of many science fiction movies (my personal favourite is *Mad Max* – not that I'd like to live in that world). We now know. Climate change is coming for all of us.

Fart Proudly

But first, let's talk about the weather. The development of complex speech as a tool for communication is a defining feature of the evolution of the human species. What were the first conversations about? My cryptocurrency is on the weather. It could be one of the defining features of us as a species. We probably do it to take the sting out of a social situation. To say we're not a threat to the person we're talking to. But also, knowing about the weather is useful. It informs us as to when to plant our crops. Or as an old Irish saying goes: 'Don't thatch a roof on a windy day.' A certain group of humans might have left Africa 200,000 years ago because they thought it was too hot, but we don't know that (yet). One human might well have turned to another and said: 'It's far too warm here, let's move somewhere it's a bit cooler.' Some might have liked the cold of winter as they moved north.

In the UK, the weather is famously a major topic of conversation (well, other than those other really compelling topics: Brexit, football and the royal family). Weather is obviously relevant to the subject of climate change, but importantly weather is not climate. Weather is the state of the atmosphere with regards to heat, cloudiness, dryness, sunshine, wind, rain, etc., while climate is the weather conditions prevailing in an area in general or over a long period. Meteorologists say: 'Climate is what you expect. Weather is what you get.' The weather might well change from day to day or week to week, and clearly it changes with the seasons. Despite these fluctuations, it is possible to build a picture of the characteristic weather patterns in different parts of the world. In Ireland, for example, we have centuries of experience that tell us that, on average, winters will be damp and cold and summers will be damp and a bit less cold. As my father used to say, you can tell it's summer in Ireland because the rain is a bit warmer. Unfortunately, this stability is no longer certain, as climates are changing at a rapid rate. 'Change' is the key word here. It may be reflected in the average temperature shifting up or down, or the amount of rainfall, or the frequency of extreme events. These persistent trends in weather are what counts when it comes to climate. And they are changing in different parts of the world, as we will see. The question is, what does this mean for the Earth and our own survival – and what can we do about it?

Edmond Halley, the important astronomer we met earlier (and no lover of fish), was also interested in climate, and published one of the first maps of winds that reliably blow east to west to the north and south of the equator in 1686. They help ships to travel west. However, Benjamin Franklin gets the credit for the first map of the Gulf Stream, the warm and fast Atlantic Ocean current that originates in the Gulf of Mexico and flows to northern Europe. Among many roles, Franklin was the first Postmaster General in the US, and mapped the Gulf Stream to help ships deliver mail to Europe more quickly. This is an early example of what would become climatology, given how important the Gulf Stream is for keeping northern Europe more temperate than it otherwise would be. It keeps mild air over north-western Europe, especially in Ireland, the UK and coastal Norway, and has been called Europe's central heating system. Major changes in climate might alter its speed or direction, which would lead to a major shift in climate in the Northern Hemisphere. It has been estimated that, were that to happen, Ireland and the UK would have winters as cold as Toronto in Canada.

Benjamin's father, Josiah, was a remarkable enough man in his own right. He was born in Northamptonshire in England in 1657 and had a total of 17 children with his two wives (who, I hasten to add, he wasn't married to at the same time). He was a soap and candle maker and emigrated to Boston in 1682. Selling soap and candles was a very lucrative business, and so Josiah became wealthy. Franklin was born in Boston in 1706, Josiah's 15th child overall, and his 10th son. This was a family that must have had robust immune genes, given the usual high level of mortality from infectious diseases, especially among children at that time.

Franklin only went to school until the age of 10. But he was an avid reader and became apprentice to his brother James as a printer. When Benjamin was 15 his brother founded the *New-England Courant*, which was one of the first American newspapers. Benjamin wrote a letter to the paper for publication, but his brother refused to publish it. He therefore wrote under the name of 'Silence Dogood', pretending to be a middle-aged widow. Luckily enough he didn't have to present himself to his brother as Mrs Dogood, as it's unlikely that a 15-year-old boy would have pulled that one off.

His letters were published and became the subject of much discussion in polite circles. He was in effect a columnist for the paper, but as far as we know his brother never found out. Franklin (as Dogood) wrote on many topics, but especially freedom of speech – a hot topic in what were still the American colonies. When his brother was jailed for three weeks in 1722 for insulting the governor of Massachusetts in print, Dogood wrote: 'Without freedom of thought there can be no such thing as wisdom and no such thing as public liberty without freedom of speech.'

At the age of 17, he ran away to Philadelphia, working for several printers. He got to know the governor of Pennsylvania, Sir William Keith, who sent him to London to acquire a new printing press with which to establish a newspaper in Philadelphia – but the plans came to nothing. He returned to Philadelphia, becoming a journalist, and eventually owned several newspapers himself. His eldest son died of smallpox. His wife had refused to have him variolated and it is likely that it was this which made them become estranged. Does that sound familiar to anyone?

Although this early example of 'vaccine' hesitancy may have been the cause of marital disharmony, Franklin's frequent extended voyages to Europe could also have played a part. It would also be fair to say that doctors using dried scabs or fluids from people with smallpox to provide protection (variolation) was a long way from the vaccines we currently use. Although the scabby mess used in variolation was effective in protecting from future infections with smallpox, it did kill 1–2 per cent of people treated – though this was much better than the 30 per cent death rate for people infected with smallpox at the time.

Much as today people might tweet, Franklin wrote many pamphlets and letters (often under pseudonyms – again, like on Twitter) which were highly popular. These were often witty and coined phrases such as 'Fish and visitors stink after three days'. One of my favourites is an essay called 'Fart Proudly'. Written in 1781, this was a criticism of various academic societies in Europe which Franklin felt were too pretentious and serious and not sufficiently concerned with practical matters ('up their own asses', as I might say). The essay discussed the way different foods affected the odour of flatulence and proposed scientific testing to get to the bottom of it. It ended by claiming that, compared to the practical applications of the essay, other sciences were 'scarcely worth a FART-HING'. Apologies for doing a Charles Babbage and ruining the joke by explaining it, but a farthing was a small coin valued at one-quarter of a penny, in use up to the end of 1960 in the UK. My father would have been a fan of Franklin, as one of his sayings to me as a child was to encourage me to fart: 'Wherever you may be / Let your wind blow free.' I'm not sure what my mother made of that. Today, the methane in the flatulence of cows is a pressing environmental concern. In fairness to the cows, they don't know that.

Franklin had a strong interest in science and is credited with inventing an impressive range of technologies, from bifocal glasses to the urinary catheter. In 1746 he began exploring electricity but was disappointed that his work resulted in 'nothing in the way of use to mankind'. He did however organize a dinner party in which he killed a turkey by electric shock and then roasted it on an electric spit, concluding that 'birds kill'd in this manner eat uncommonly tender'.

He famously flew a kite in a storm to prove that lightning was electricity. These experiments gave rise to the invention of the lightning rod. He made sure to stand on an insulator under a roof to avoid electric shock. Sadly, others who repeated his experiment didn't do this and weren't so lucky.

While on a visit to England he heard that it took ships carrying mail several weeks longer to reach New York than Newport, Rhode Island. He asked his cousin who was a Nantucket whaler why (illustrating a key trait for a scientist: always be wondering) and was told that the faster passage avoided strong mid-ocean currents. Mariners had known about these for generations, but no one had systematically studied them. With the help of ship captains, Franklin mapped what he called the Gulf Stream. British sea captains followed this map, avoiding the currents and shortening their passage to and from the US. This is the start of the science of oceanography, an important part of climate science research.

Cloud Atlas

The scientific study of climate really begins with Wladimir Köppen, who classified the main types of climate. This is known as the Köppen climate classification, which is nicely alliterative. It is still in use. There are five main climate types: tropical, dry, temperate, continental and polar. Köppen based these on the types of vegetation growing in a region. The link to plant life is useful in predicting climate change, since if the plant life is changing, it is likely that the climate is changing too, which as we'll read later is indeed the case.

Köppen was born in 1846 in St Petersburg. His grandfather was a German doctor invited to Russia by Catherine the Great to improve sanitation. Catherine had kept up with developments in clean water to prevent the spread of infectious diseases like cholera, and wanted the best for the people of Russia. She was great. Köppen's grandfather went on to become the personal physician to the tsar. Köppen's father was a noted geographer, who obviously influenced his son. Köppen went to school in Crimea, and it was while journeying there from St Petersburg (a distance of over 2,000 kilometres) that he noticed the diversity in plant life. In 1870, he was awarded a PhD at the University of Leipzig on the effect of temperature on plant growth, eventually becoming the head of the marine meteorology department in Hamburg. Like many scientists, myself included, his interest began with a question: why do different plants grow in different places?

Köppen is the founder of climatology and meteorology. He published his first climate classifications in 1884, and went on to publish over 500 papers – including in 1890 the first cloud atlas, a book describing the main cloud types. It was used by meteorologists to help classify clouds, which was useful for weather forecasting. The book was a great success, and why wouldn't it be? Everyone loves clouds. It featured state-of-the-art photographs, too, which people enjoyed looking at.

Köppen also had an interest in palaeoclimatology, the study of ancient climates, and with his son-in-law Alfred Wegener – who you may remember was the first to suggest continental drift as an explanation for continent formation – published an important book on that topic. They must have had great discussions around the family dinner table. Theirs was a key book on the subject, providing important evidence for the ice ages – the evidence in part relying on the physical features that glaciers create such as moraines, U-shaped valleys and eskers, as we read about in chapter 3.

Köppen was also a promoter of world peace and Esperanto, a constructed language designed to unite the people of the world – which sadly failed in that goal, but not for want of Köppen trying. He could speak Esperanto, and between 1868 and 1939 he translated many of his works into that language. We don't know if anyone ever read them.

As the twentieth century rolled around, climatologists began collecting detailed data on each climate type, including temperature fluctuations but also changes in atmospheric pressure and wind speeds. This led to descriptions of climate variability. Various 'oscillations' were described, including the El Niño–Southern Oscillation – an ocean-atmosphere phenomenon in the Pacific Ocean that is responsible for most of the global variability in temperature, and has a cycle of between two and seven years. Oscillations in wind speeds and sea-surface temperature occur, with the warming phase of the sea known as El Niño, and the cooling phase La Niña.

There are three other major oscillations in wind speed and sea temperature: the North Atlantic Oscillation (driven by the aforementioned Gulf Stream), the Madden–Julian Oscillation and the Interdecadal Pacific Oscillation, which has a cycle of 15–30 years. These oscillations have been remarkably stable over decades of observation, but lately things are beginning to change, providing more evidence that we are in an era of climate change.

Climate change is defined as new weather patterns emerging in a given climate that remain in place for an extended period of time. We have accurate measurements of weather conditions over the last 200 years or so in most climate zones, thanks to thermometers (which measure temperature), barometers (which measure pressure), anemometers (which measure wind speed) and various other -ometers. As with every branch of science, advances happen because of the invention of instruments or techniques to accurately observe something, and climatology is no different.

Solar Ovens and Greenhouses

To get a real handle on changes in climate we need to go back further in time, to before the Industrial Revolution and the associated expansion of humanity. Various analyses (for example, using ice cores from Antarctica) have revealed that the Earth's climate underwent major changes in the past, even when we humans weren't there. Ice cores are long tubes of ice removed from an ice sheet or mountain glacier. They are laid down year after year, and so analysing a layer gives a record of the climate of that year.

Climate is mainly governed by the energy arriving from the sun, but also how much energy is given off into outer space. When the incoming energy is greater than the outgoing, warming occurs; whereas when it's the opposite, cooling occurs. Both have happened throughout the Earth's history and have mainly been governed by the strength of the sun and changes in the Earth's orbit – say, if the Earth is tilted slightly closer to the sun or slightly further away. Periods of warming have led to higher sea levels because of the melting of ice and more extreme weather, and periods of cooling have led to ice ages, which have locked up water as ice and reduced sea levels and led to the sculpting of the Earth's terrain by glaciers.

That was something that always interested me greatly as a

child and may in fact have been the reason for me becoming a scientist. I grew up in County Wicklow, and when I heard that the mountains had been carved out by huge glaciers 10,000 years ago it did something to me. Not quite sure what. It just seemed so majestic and magical and wonderful that the beautiful Valley of Glendalough was U-shaped (as opposed to V-shaped) because a big moving glacier had carved it out. The glacier would have ground through the V-shaped valley, smoothing out the valley floor into a U-shape. Once the climate warmed, all the ice melted, leaving behind these remarkable and beautiful features.

That type of climate change wasn't caused by humans. But what concerns us today is climate change that *is* being caused by human activity. And that of course is where controversies come in. Recent studies have shown that 98.7 per cent of climate change scientists agree that it is human activity that is causing climate change. As we've already read, the bottom line for a complex scientific question often comes down to a consensus, because there may well be some evidence against what is being proposed. And the consensus should leave us in no doubt.

The greenhouse effect was first proposed by Joseph Fourier. He was a French mathematician born in 1768, and in the 1820s he calculated that given the amount of energy coming from the sun, a planet the size of the Earth should be much cooler than it is. He proposed that the Earth's atmosphere might act as an insulator. His publications referred to experimental data from Horace de Saussure, the inventor of the first solar oven, which used a vase over which glass panes were placed with air in between. When illuminated by the sun, the temperature would be elevated in the vase. Fourier concluded that the atmosphere might act like the panes of glass, similarly elevating the temperature of the Earth. We now define the greenhouse effect as the means by which the atmosphere traps heat from the sun, raising the Earth's temperature. But what about the evidence?

Foote's Boots and Shoes

Eunice Foote was born in 1819 in Connecticut. Her father was Isaac Newton. Foote's father was a distant relative of the famous Newton of falling-apple fame – and from my excellent powers of deduction, so was Foote, obviously enough. She was lucky enough to be taught by Amos Eaton, who was a key science educator in the US. The school Eaton founded in New York in 1824, the Rensselaer School, focused on 'the application of science to the common purposes of life'. She studied chemistry, which was unusual for women at that time, and she was especially influenced by a textbook written by Almira Hart Lincoln Phelps, *The School Girls' Rebellion* – a great title if ever there was one. Phelps was a pioneer of women in science, and the third female member of the American Association for the Advancement of Science.

Foote was an early campaigner for the rights of women, and she was an early signatory of the 1848 Declaration of Sentiments, which demanded equality with men in social status and legal rights – including the right to vote. The first experiments that she reported on involved studying the interaction of sunlight with different gases. She used a system of glass tubes with thermometers to test the effect of various gases on temperature. She saw that CO_2 trapped the most heat and wrote that 'an atmosphere of that gas would give our Earth a higher temperature.'

Although women were not restricted from presenting at the American Association for the Advancement of Science's conferences, Foote's work was presented by a man, John Henry of the Smithsonian Institute. Before Henry introduced her findings, he said, 'Science was of no country and no sex.' Foote's work was published under her own name in 1856, followed by a publication in the same edition of the journal by her husband, Elisha, who was a mathematician of some note. In an example of something that will resonate with anyone who has had their work summarized or edited (and I mean me), the summaries in European journals omitted her findings on CO_2, and in one, the *Edinburgh New Philosophical Journal*, her husband was given the credit. She must have been spitting nails when she heard that. In 1856, *Scientific American* praised her in an article called 'Scientific Ladies', emphasizing how her findings were backed by experiments. Before her work it was only theorized that a gas like CO_2 would trap heat. The article concluded: 'this we are happy to say has been done by a lady.'

Foote was also an inventor, and in 1864 made a new type of paper-making machine which produced a product of higher quality. She had a patent on 'filling for soles of boots and shoes' that 'prevented the squeaking' – which for some reason was especially annoying to Foote, whose name was never more appropriate for such an invention.

Three years later, in a study foreshadowed by Foote's work – about which, as far as we know, he wasn't aware – Irish scientist John Tyndall reported on how various gases can trap and emit infrared thermal radiation. Tyndall had trained to be a chemist under none other than Robert Bunsen, inventor of the famous Bunsen burner. He was an avid mountain climber and led one of the first teams to the top of the Matterhorn. His interest in mountains led him to study glaciers, and he has glaciers named after him in Chile and Colorado, and mountains named after him in California and Tasmania. His work on glaciers led him to study the heating effect of sunlight – in particular the transmission and absorbance of heat by different gases. He knew there had been ice ages in the past, including the one that led to the Wicklow Mountains being carved out, and wondered how they had come to an end.

In 1859 he described what eventually became known as the greenhouse effect, writing 'the atmosphere admits of the entrance of solar heat; but checks its exit, and the result is a tendency to accumulate heat at the surface of the planet.' Seemingly unaware of Foote's work, Tyndall speculated on how different levels of gases in the atmosphere might lead to ice ages.

Tyndall was an advocate for communicating science to the public. He gave many public lectures, especially at the Royal Institution in London, and went on several highly lucrative lecture tours of the US, donating the profits to science education. He said of the occupation of teacher: 'I do not know a higher, nobler and more blessed calling.' At the age of 55 he married Louisa Hamilton, the 30-year-old daughter of an MP. In his later years he suffered from insomnia and took chloral hydrate as a remedy. Sadly, at the age of 73 he overdosed on it and died. Unfortunately, the overdose was given by his wife and his last words were: 'My darling, you have killed your John.' Not the happiest of endings.

Not long after Tyndall's death, Swedish scientist Svante Arrhenius published the first climate model, theorizing that a halving of carbon dioxide in the atmosphere could initiate an ice age. Like so many of the scientists we have mentioned,

Arrhenius had been something of a child prodigy. He was born near Uppsala in 1859. He learned to read at the age of three, apparently without the help of his parents. He studied chemistry at the University of Uppsala, but fell out with his tutor, Per Teodor Cleve, and moved to Stockholm. Arrhenius's dissertation on electrolytes wasn't well received, possibly because Cleve was one of the examiners, and he barely scraped a pass. But the work in the thesis led to the basis of his 1903 Nobel Prize. In 1896 he was the first to conclude that CO_2 emissions caused by humans from fossil fuel burning were large enough to cause global warming. He had built on the work of Foote, Fourier and Tyndall. This is how science works – sometimes incrementally, sometimes in quick bursts, but always backed by evidence.

The term 'greenhouse effect' as applied to global warming was coined by Nils Gustaf Ekholm, a lifelong friend and colleague of Arrhenius, in a presentation to the Royal Meteorological Society in 1901: 'The atmosphere plays a very important part of a double character as to the temperature at the earth's surface, of which the one was first pointed out by Fourier, the other by Tyndall. Firstly, the atmosphere may act like glass of a green-house, letting through the light rays of the sun relatively easily, and absorbing a great part of the dark rays emitted from the ground, and it thereby may raise the mean temperature of earth's surface.' Sadly, no mention of Foote. But a great analogy to explain in simple terms what was happening.

Scientists initially rejected the idea of the greenhouse effect. Not for the first time was a radical idea being rejected by the consensus view. Nor the last. In 1938, climatologist Guy Stewart Callendar published evidence that the Earth's climate was warming and CO_2 levels were increasing, but his calculations were questioned. For a while, this was called the Callendar Effect, and Callendar thought it might be a good thing, delaying a 'return of the deadly glaciers'.

Callendar had compiled measurements of temperature from the nineteenth century and correlated them with measurements of atmospheric CO_2. He concluded that over 50 years the global land temperature had increased, and proposed that this increase was caused by the increase in CO_2. The estimates were accurate – especially impressive as they were done without a computer by a man who was partially blinded as a child after his brother stuck a pin in his eye.

The most eminent meteorologist at that time, George Simpson, said that the correlation must be merely a coincidence. Yet again, the issue of correlation and causation, but in this case Simpson should have realized that the rise in CO_2 would cause an increase in temperature. In a great example of dogged scientific perseverance, Callendar's work over the next 20 years convinced many scientists to study the link in more detail. His publications in the 1940s and 1950s had one message: global warming was happening because of CO_2 emissions. In 1958, Charles Keeling published data from the Mauna Loa Observatory that agreed with Callendar, which proved very influential in the debate on human influence – the rise in CO_2 became known as the Keeling Curve.

But warnings from scientists went largely unheeded. It looks like many scientists were in denial. Multiple lines of evidence, arrived at via different approaches, all pointed relentlessly to one conclusion: the climate system had been warming since the industrial era began, and the two were linked.

In the mid-1960s, an analysis of deep-sea cores by Cesare Emiliani and of ancient corals by Wallace Broecker came to an important conclusion concerning ice ages. Emiliani discovered from the ice cores that there had been several ice ages over 500,000 years and that they were cyclical. Broecker worked on how the circulation of the oceans was an important determinant of climate. Both scientists concluded that climate could be sensitive to small changes and could flip quite quickly from one mode into another. Broecker used the term 'global warming' in 1975, and continued up to his death in 2019 to urge scientists to study solutions to the problem.

Project Iceworm

The rise of computers allowed complex modelling to be carried out that simulated the impact of different levels of CO_2 and other greenhouse gases, and one model concluded that a doubling in CO_2 levels would result in a two degrees Celsius rise in global temperatures. To put this in perspective, the Mauna Loa measurements have shown that between 1960 and 2022, CO_2 levels rose from 315ppm to 415ppm, which is heading towards the doubling that will cause a rise in global temperature.

Evidence for global warming from the flowering time of plants has only grown. One study examined 590 species of flowering plants over the last 30 years, including herbaceous plants, cacti, shrubs and trees. A shift to earlier flowering time was associated with increased temperature. In a great example of citizen science, which involves members of the public collecting data, more than 400,000 observations were made of 406 plant species. Comparing 1987 to 2019, average first flowering was a month earlier than the average between 1753 and 1986. This has worried scientists, because the flowering might be out of sync with when insects arrive to pollinate flowers. The insects may therefore not be able to feed, which in turn could have a knock-on effect for the birds that feed off the insects. Such analysis therefore isn't only important for providing evidence for climate change, but also for the possible consequences of that change.

Compelling evidence has also come from the analysis of ice cores, the longest of which stretch back 800,000 years. The properties of the ice and the material trapped in it can be used to reconstruct the local climate over the age range of the core. Air trapped in tiny bubbles can be analysed for CO_2 levels. Pollen can be collected for information on the types of plants that grew at particular points in time. Volcanic events can also leave a signature. All kinds of elaborate technologies are being used to assess ice cores, from mass spectrometers and gas chromatographs (which can identify different molecules) to scanning electron microscopes (which can observe tiny structures in the ice core like pollen grains or types of dust, based on their shape). Effectively these are time machines; you're looking back in time when you assess an ice core.

Antarctic ice cores have revealed that the concentration of CO_2 was stable in the atmosphere over the last thousand years, until the early nineteenth century. It then started to rise, and its concentration is now almost 50 per cent higher than it was before the Industrial Revolution. Analysing the isotopes of carbon (meaning their radioactivity, which gives an idea of where the carbon comes from) has shown that the recent increase is due to the emission of CO_2 from fossil fuels. Looking back, using our ice time machine, although CO_2 levels did show very slow periodic oscillations (from about 170 ppm to 290 ppm), this rate of rise is unlike anything that has happened over the past 800,000 years. For example, there was a rise from 200 ppm to 280 ppm between 450,000 to 400,000 years ago, i.e. an 80 ppm rise over 50,000 years. We've seen an 80 ppm rise just in the last 45 years.

Meanwhile, methane – another key greenhouse gas – has shown an unprecedented rise in the past 200 years, coming largely from rice fields and farm animals. The analysis of the Greenland ice cores examining oxygen isotopes to estimate temperature is also providing some of the most compelling evidence for recent global warming. Temperatures in Greenland are warmer than at any time in the past 2,000 years. The way things are going, by 2050 it will be the warmest in Greenland since the last interglacial period of 125,000 years ago.

One interesting analysis concerns a sample taken in 1966 that was a mile long. At the bottom of it was a soil sample which was stored, unanalysed, in a freezer in a Danish lab. The ice core was extracted during the (wait for it) Cold War during a military mission called Project Iceworm. This was a plan to hide nuclear weapons from the Soviets far to the north, in the Greenland icefields. Whoever had that idea must have been mad, as digging in the permafrost is difficult, although hiding the weapons in a glacier is probably more feasible since the ice could first be melted to get the weapons in, as long as it's not a transparent glacier. A place called Camp Century was built but didn't last long, as the snow and ice began crushing the buildings and the underground tunnels that were finally dug.

Samples of the ice core were kept at the University of Buffalo before being transferred to Denmark, where they were forgotten about. Then someone rummaging through the freezers found them. Microscopic analysis revealed plant material in the thawed ice, indicating that during a period of thawing, a tundra ecosystem grew in north-western

Greenland. The analysis showed that less than 1.1 million years ago there was no ice sheet at all in Greenland. It had melted to the ground. At that time, the oceans would have been three metres higher. If London, Boston and Shanghai had existed at that time at their current heights, all would have been underwater.

Retrieving the ice core sample was fortuitous, as if the core had been analysed in 1963, given the techniques that were available at that time, not a lot of information would have been found. What would happen if the ice in Greenland were to melt again because of global warming? Sea levels would rise more than six metres, redrawing all the coastlines of the world.

A Damning Indictment

In 1988, the Intergovernmental Panel on Climate Change (IPCC) was established. This is a body of the United Nations which aims to advance knowledge on human-induced climate change and to advise governments accordingly. The IPCC has produced six reports, and each successive report provides more and more evidence of climate change and the negative consequences for life on Earth.

The sixth report, published in 2021, is worth summarizing, as it is the most important assessment of what's going to happen to the Earth in the coming years, so we'd better listen up. The first part concerns the 'Physical Science Basis', and 234 scientists from 66 countries contributed. The report builds on more than 14,000 scientific papers, and 195 governments signed up to its conclusions. It states that the only way to avoid warming of 1.5–2 degrees Celsius (which is in the same range as the warning issued by Arrhenius over 100 years ago, and was agreed by 195 governments in 2015 as part of the Paris Agreement) is for a massive and immediate cut in greenhouse gas emissions. If this can't be achieved, then the consequences for the Earth will be severe – especially for the poorest regions, many of which are at risk from flooding and crop failures.

A key concern is that the Earth will reach a tipping point, beyond which certain impacts can no longer be avoided even if temperatures are reduced. One example is if the Gulf Stream were to halt because of a change in the salinity of the Atlantic Ocean. This would trigger major climate change in Europe and North America, and this would be almost impossible to reverse.

The latest IPCC report highlights two particular concerns for humans: famines because of crop failure, and flooding of low-lying regions. Strategies such as the mass migration of people, or what is called 'managed retreat' from coastal areas are discussed. More recently it has been reported that the incidence of many infectious diseases is rising. A study examined 375 infectious diseases and found that 218 of them had got worse in terms of how common they are and their severity. One reason for this is that some diseases can be carried by insects like mosquitoes, and if they can live in new places because of increasing temperatures they will bring the diseases there.

The second part of the IPCC report was published on 28 February 2022 and is even more pointed in its concerns. It opens with the disturbing prediction that at least 3.3 billion people, or about 40 per cent of the world's population, now fall into the most serious category of 'highly vulnerable', with the worst effects again being felt by the developing world. It predicts that if greenhouse gas emissions continue on their current path, Africa will lose 30 per cent of its useable land for growing maize and 50 per cent of its land for beans. One billion people face flooding because of rising sea levels. The report identifies 127 different negative impacts of climate change, some of which would be irreversible.

These first two parts of the report focused on understanding and outlining the scale of the problem, and struck fear into many – with the Secretary General of the UN, António Guterres, calling it 'an atlas of human suffering and a damning indictment of failed climate leadership'.

The IPCC issued the third part of the sixth report on 4 April 2022, which rather mercifully contains a range of mitigation measures. These are all achievable if we have the will to do them. There's a major emphasis on global cooperation and CO_2 removal, including promoting forestry (trees remove CO_2 from the air), city planning for public transport, replacing fossil fuels with renewable energy sources to power electric vehicles, and a whole host of other potential solutions. There are also simpler strategies. If cows are fed seaweed as part of their diet, it would reduce methane emissions by 80–95 per cent. Reducing cow burps and bringing hope.

A range of technological solutions are also being explored. This area is termed geoengineering, and various strategies to deflect sunlight away from the Earth are being investigated. One of these is intriguing. It's called marine cloud brightening and involves spraying sea water into the air to make clouds whiter, thereby reflecting more sunlight, like a shinier mirror. The only problem is it comes at a cost, as energy on a huge scale is needed to spray the water into the sky. Another involves floating solar installations, which are solar panels floating on existing hydroelectric dams. They generate power during the day while hydroelectric power is created at night. There are so many reservoirs that they could be deployed in. Imagine – all reservoirs covered in solar panels.

There has been a major increase in public awareness of the data showing that human activity is affecting climate, and many people are taking personal action by, for example, recycling, better insulating their homes, using bicycles, taking public transport, buying electric vehicles, and voting for political action that can reduce our climate impacts. Dissent from climate change sceptics appears to be lessening, but still 3–10 per cent of the US, UK and Australian populations deny that climate change is happening. But we should be encouraged that more than 95 per cent take it seriously.

Greta Says We Cannot Make It Without Science

The Swedish environmental activist Greta Thunberg has done more to promote debate on the issue of climate change than any scientist. We can all learn something from Thunberg as to how to get the message across. She is following in the tradition set by her Swedish compatriots Arrhenius and Ekholm.

Thunberg was born on 3 January 2003, the daughter of opera singer Malena Ernman and actor Svante Thunberg. She first heard of climate change at the age of eight and wondered why it was being ignored. For about two years she challenged her parents to lower the family's carbon footprint. Her family became vegan and stopped flying, which meant her mother giving up her international singing career.

At the age of 15 Thunberg won a climate-change essay competition held by Swedish newspaper *Svenska Dagbladet* in which she wrote 'I want to feel safe. How can I feel safe when I know we are in the greatest crisis in human history.' After Sweden's hottest summer in 262 years, she started spending her school days outside the Swedish parliament, calling for stronger action on climate change. She demanded that the Swedish government reduce carbon emissions in accordance with the Paris Agreement.

She posted a photo of her first strike on Instagram and Twitter, and on the second day she was joined by several other protestors. Her posts began to go viral. After October 2018 she took part in demonstrations throughout Europe, making many speeches, and inspired school students all over the world to take part in strikes. She attended the 2018 United Nations Climate Change Conference (COP24). To avoid flying, she sailed to North America from Plymouth in the UK. Then the conference was moved to Madrid, so she sent out a message on social media asking for a ride across the Atlantic. She hitched a ride on board a catamaran that brought her to Lisbon.

In her speech at the conference she said: 'My message to the Americans is the same as to everyone – that is to unite behind the science and to act on the science.' She accused the world leaders present of being 'not mature enough to tell it like it is'. In January 2019 she gave a speech at the World Economic Forum in which she said 'our house is on fire', and in 2020 said: 'Our house is still on fire.' She has recently stated that climate experts are still not being listened to despite the COVID-19 pandemic highlighting the importance of using science to address important issues. She said that the COVID-19 crisis 'shone a light' on how 'we cannot make it without science'.

Thunberg is leading the way. But the key question remains.

Given how smart we are, with all the great science and technologies at our disposal, can we save the planet that we've been destroying? The scientific consensus says we must act, and fast. Marie Skłodowska Curie famously said: 'Nothing in life is to be feared. It is only to be understood. Now is the time to understand more so that we may fear less.' We understand about climate change, and have done for hundreds of years. We now need to use that knowledge to stop it and get rid of the fear. Otherwise, what will all this science have been for?

11. I Still Haven't Found What I'm Looking For

I only have a vague memory of my grandmother, Agnes Bourke O'Neill. She died when I was three years old. But two memories stand out. The first is me sitting in her garden on what I still remember as intensely bright green grass, beside some geraniums. I was amazed at how green the grass was and the heady scent coming off the flowers. The second is even more vivid. I was sitting on the floor in her living room, and she was holding a small feather. She let it drop, and I remember it floating slowly to the floor. She picked it up, and let it drop again. I was transfixed and filled with awe at the sight. Was that simple physics experiment the start of me as a scientist?

As children, we are filled with wonder as we discover the world. And we'd better make sure we encourage our young people to keep wondering. We should especially encourage young people who think a bit differently to become scientists. As you'll hopefully have realized, some of the best scientists were wackos. We need more wackos. As Carl Sagan, that great science communicator, famously said, 'Something somewhere is waiting to be known.'

Think about that for a minute. Think of all the things we've found out already. And there's so much more still to be discovered.

Robert Boyle grappled with this and predicted what might be discovered in the future, and 350 years ago, he drew up a list of things he hoped could be achieved through science.

It's intriguing to examine that list now and see how far we've come. He listed the following, and under each I've put the progress we've made:

Prolongation of life

This has been achieved to some extent: modern medicine and lifestyle changes are allowing us to live longer. The average lifespan in Robert Boyle's time was 35 years, so we've more than doubled that (to 82 years).

Recovery of youth

Sadly, this is not yet possible, unless you count Botox or hair transplants. Scientists have yet to come up with a way to reverse ageing.

Art of flying

Achieved – as we read about in chapter 2: from balloons to planes to rockets.

Art of continuing long underwater

Achieved too – with the invention of the submarine and scuba gear.

Cure of diseases at a distance or at
least by transplantation

It's not really clear what Boyle meant by 'at a distance', but there are now robots that can perform surgery and radiotherapy can be used to treat certain cancers. Transplantation has been achieved for several organs.

Cure of wounds at a distance

This has also been achieved to some extent, with the discovery of drugs that can promote wound healing such as pentoxifylline, which works by increasing blood flow to the wound which then promotes repair.

Transmutation of metals

Radioactive decay can change one element into another, but we haven't quite managed to change lead into gold.

The making of armour light

This has been achieved in the form of Kevlar, a strong, heat-resistant synthetic fibre made from a chemical called poly-para-phenylene terephthalamide (Kevlar is easier to say). Kevlar can protect against stabs – and when made of 15 layers, even bullets.

Finding longitude

This was achieved when in 1730 John Harrison invented his clock, which could measure time with remarkable precision – and we have GPS now.

Potent drugs to exalt imagination

This has been achieved with the discovery of drugs such as LSD – but trippy, man!

A ship to sail with all winds

Achieved – with the invention of engines to drive ships.

*

So, we've made amazing progress. Boyle would be delighted because the scientific method he advocated for has been put to great use. But what might happen next? Several predictions were made in the TV show *The Simpsons* that have come true, notably the GoPro camera, autocorrect (even the errors: a bully writes down in an electronic diary 'Beat up Martin', which is autocorrected to 'Eat up, Martha'), virtual reality glasses, videochatting and the smartwatch. *The Simpsons* is indeed the font of all human knowledge.

Here is my own list of what science might achieve in the future. I wonder how many of these will be achieved in my lifetime and beyond.

Figuring out how the mind works

As I wrote about in the chapter on the brain, this is still a mystery. We have no notion of how consciousness works, for example. If we did, I'd tell you.

Discovering why we age and being able to slow it down

This question persists from Boyle's time. Ageing is a part of life and it's still not clear what the underlying mechanism is. A lot of effort is going into trying to understand why we age, and there has been some progress.

The conquering of all the major diseases that afflict us, with equitable access to healthcare for all

This will most likely involve gene editing to correct the faulty genes that cause so many diseases. As we saw in chapter 6, this might involve the DNA editing technology called CRISPR.

It might also involve the invention of vaccines for many diseases, including cancer, or better ways to deliver therapies to where you want them to go – with for example nanoparticles, an approach I am researching myself.

The halting of species becoming extinct and the restoration of those that already have

This might involve the retrieval of DNA from the extinct animal, which could then be used to somehow fertilize an egg which might then develop. If we want them back – we might be wise to leave out the dinosaurs.

Effective 3D printing of everything we need – just like the replicator in *Star Trek*

This will mean an end to money, as why would money be needed if the only transactions involved buying a 3D printer and its supplies – as all the key things can be printed at the flip of a switch? The technology for 3D printing continues to improve.

A renewable and safe source of energy that will provide us with power for all our needs without contributing to global warming

This is a very active area of research for the obvious reason that the way we currently extract energy is killing the planet. One goal is to achieve fusion, where more energy is released than was put in. There was progress in 2022, when two isotopes of hydrogen – deuterium and tritium – were fused in an experiment that was hailed as a big success, with the warning that turning this into a viable source of energy is still a long way off. But still, progress nonetheless.

All vehicles, (planes, trains, automobiles and beyond) becoming driverless

In the case of automobiles, this will mean an end to traffic accidents and traffic jams. Advances are being made in this area too, with artificial intelligence being deployed as well as a range of computer and imaging technologies.

Rapid global travel using a renewable energy source

Imagine getting to Australia in one hour. Ideas like the hyperloop, which is a high-speed transportation system for people and goods, are being explored. It involves magnetic propulsion in an almost frictionless tube.

Space travel for the fun of it

The Artemis missions plan to put humans on the moon by 2025, building a lunar gateway which will act as a jump-off point for a mission to Mars. We may need to go there if the Earth is dying, but hopefully not.

The discovery of extraterrestrial life

Surely only a matter of time. I can't wait to see what it looks like.

It's a sign of how fast science and technology are evolving that I am confident that many of these will become reality within the next 50 years – and most within a century. Advances in artificial intelligence will hopefully speed up progress but not lead to our demise at the hands of machines cleverer than us (another sci-fi trope). We can catch up for a pint to see who was most correct in 2073. The loser buys the drinks!

The goal of science is to make discoveries, and then for those discoveries to be used by society to benefit all life on Earth.

Think of all the things we've found out already. When my grandmother Agnes Bourke O'Neill was born in 1893 (more than a lifetime ago, but a drop in the ocean of how long it's been since the Earth formed) we didn't know how big the universe is, about black holes, or about viruses or DNA or treatments for so many diseases that afflict us, and there was no notion of computers and the internet. I wonder if Agnes knew that with her demonstration of the aerodynamics of a feather she had inspired her grandson to embark on a lifetime of science, making discoveries about the immune system that will hopefully one day lead to better treatments for autoimmune and auto-inflammatory diseases?

Scientists have been participants in the best reality show of all time. With all the highs, lows, failures, conflicts, bust-ups and strange personalities. Only, this one has given us so much more than entertainment. The scientists have also given us hope.

Further Reading

The following is a list of some of the sources I used in the writing of this book, which I think you'll find informative and entertaining – provided you enjoyed this book in the first place. Keep on reading!

Introduction

Paolo Rossi, *Francis Bacon: From Magic to Science*, Routledge, 2013.

Gyeong-Geon Lee and Hun-Gi Hong, 'John Amos Comenius as the Prophet of Modern Ideas in Science Education: In the Light of Pansophia', *History of Education*, vol. 50, no. 1, 2021.

1. City of Stars

Geraldine Stout and Matthew Stout, *Newgrange*, Cork University Press, 2008.

Arthur Koestler, *The Sleepwalkers: A History of Man's Changing Vision of the Universe*, Penguin, 1989.

Chris Gosden, *Magic: A History. From Alchemy to Witchcraft, from the Ice Age to the Present*, Farrar, Straus and Giroux, 2020.

George Johnson, *Miss Leavitt's Stars: The Untold Story of the Woman Who Discovered How to Measure the Universe*, W. W. Norton and Company, 2006.

Mabel Armstrong, *Women Astronomers: Reaching for the Stars*, Stone Pine Press, 2008.

Megan Borgert-Spaniol, *Jocelyn Bell Burnell: Discovering Pulsars*, Abdo Publishing, 2017.

2. Build a Rocket, Boys and Girls

Alberto Borgo and Ester Tomè, *The Montgolfier Brothers*, SASSI, 2018.

R. G. Grant, *Flight: The Complete History of Aviation*, Dorling Kindersley, 2022.

NASA, 'Brief History of Rockets': https://grc.nasa.gov/www/k-12/TRC/Rockets/history_of_rockets.html.

Elizabeth Howell, 'The Story of NASA's Real "Hidden Figures"', *Scientific American*, 2017, https://www.scientificamerican.com/article/the-story-of-nasas-real-ldquo-hidden-figures-rdquo/.

3. Get Your Rocks Off

Kieran D. O'Hara, *A Brief History of Geology*, Cambridge University Press, 2018.

Cathy Newman, 'The Forgotten Fossil Hunter Who Transformed Britain's Jurassic Coast', *National Geographic*, 29 March 2021.

American Museum of Natural History, 'Inge Lehmann: Discoverer of the Earth's Inner Core', https://www.amnh.org/learn-teach/curriculum-collections/earth-inside-and-out/inge-lehmann-discoverer-of-the-earth-s-inner-core.

Richard B. Alley, *The Two-Mile Time Machine: Ice Cores, Abrupt Climate Change, and Our Future*, Princeton University Press, 2014.

4. Better Living through Chemistry

The Chemistry Book, Dorling Kindersley, 2022.

Flora Masson, *Robert Boyle: A Biography*, Leopold Classic Library, 2016.

Nell Walker, *Marie Curie*, DK Life Stories, Dorling Kindersley Children, 2022.

Yolanda Ridge, *CRISPR: A Powerful Way to Change DNA*, Annick Press, 2020.

5. Living It Up

Jared Diamond, *Guns, Germs and Steel: A Short History of Everybody for the Last 13,000 Years*, Vintage, 1998.

Angela Stienne, *Mummified: The Stories behind Egyptian Mummies in Museums*, Manchester University Press, 2022.

Jude Piesse, *The Ghost in the Garden: In Search of Darwin's Lost Garden*, Scribe UK, 2021.

Nick Lane, *Transformer: The Deep Chemistry of Life and Death*, W. W. Norton and Company, 2022.

6. Message in a Bottle

R. Dahm, 'Friedrich Miescher and the Discovery of DNA', *Developmental Biology*, vol. 278, no. 2, 2005, pp. 274–88.

Luna, 'The History of DNA', 2019, https://www.lunadna.com/history-of-dna.

Heather Dawes, 'The Quiet Revolution', *Current Biology*, vol. 14, no. 15, 2004.

Brenda Maddox, *Rosalind Franklin: The Dark Lady of DNA*, HarperCollins, 2003.

Kary Mullis, *Dancing Naked in the Mind Field*, Vintage, 2000.

7. Medical Love Song

Nancy Duin and Jenny Sutcliffe, *A History of Medicine: From Prehistory to the Year 2020*, Barnes and Noble Books, 1992.

K. Codell Carter and Barbara R. Carter, *Childbed Fever: A Scientific Biography of Ignaz Semmelweis*, Transaction Publishers, 2005.

Annie Matheson, *Florence Nightingale: A Biography*, Legare Street Press, 2022.

Thomas D. Brock, *Robert Koch: A Life in Medicine and Bacteriology*, ASM Press, 1998.

World Health Organization, 'A Brief History of Vaccination', http://www.who.int/news-room/spotlight/history-of-vaccination/a-brief-history-of-vaccination.

8. Where Is My Mind?

Desmond M. Clarke, *Descartes: A Biography*, Cambridge University Press, 2012.

Muriel Spark, *Mary Shelley: A Biography*, Carcanet, 2013.

John Fleischman, *Phineas Gage: A Gruesome but True Story About Brain Science*, Clarion Books, 2013.

Ralph A. Bradshaw, 'Rita Levi-Montalcini (1909–2012)', *Nature*, vol. 493, 2013, p. 306.

Karl Deisseroth, *Connections: A Story of Human Feeling*, Viking, 2021.

9. Paranoid Android

Rachel Ignotofsky, *The History of the Computer: People, Inventions, and Technology that Changed Our World*, Ten Speed Press, 2022.

Desmond McHale, *The Life and Work of George Boole: A Prelude to the Digital Age*, Cork University Press, 2014.

History-Computer, 'Ada Lovelace – Complete Biography, History and Inventions', https://history-computer.com/ada-lovelace-complete-biography/.

History-Computer, 'Alan Turing – Complete Biography, History and Inventions', https://history-computer.com/alan-turing-complete-biography/.

Walter Isaacson, *Steve Jobs*, Abacus, 2015.

Oliver J. Rich, *NFTs for Beginners: Making Money with Non-Fungible Tokens*, n.p., 2021.

10. A Hard Rain's Gonna Fall

Wladimir Köppen biography: https://www.howold.co/person/wladimir-koppen/biography.

Stuart Mathieson, 'Eunice Foote, John Tyndall and the Greenhouse Effect', *The Irish Times*, 30 Aug 2021.

James Rodger Fleming, *The Callendar Effect: The Life and Work of Guy Stewart Callendar (1898–1964)*, American Meteorological Society, 2009.

IPCC 2022 report summary: https://climate.selectra.com/en/news/ipcc-report-2022.

Greta Thunberg, *The Climate Book: Greta Thunberg*, Allen Lane, 2022.

Acknowledgements

A big thanks to Max Edwards, who connected me to Con Brown at Viking Penguin Random House, and helped the three of us decide that I should have the temerity to write a new history of science. HUGE thanks to Con for superb editing. Like Andy Gearing, he deserves an asterisk. Also thanks to Greg Clowes, Gemma Wain and Natalie Wall for lots of excellent suggestions. Thanks to Hermann Schobesberger, who read the entire manuscript and provided important input, and to John Kelly and Damien Woods, who made key comments on specific chapters. Finally, thanks again to Andy, who is the main reason why this book will make you laugh, even though we deleted a photograph of him, and as ever to Marg, Stevie and Sam without whom.

Index